Volker Baur

White Matter Connectivity in the Brain

Volker Baur

White Matter Connectivity in the Brain

Alterations in Social Anxiety Disorder and Links to General Anxiety-Related Mechanisms

Südwestdeutscher Verlag für Hochschulschriften

Impressum / Imprint

Bibliografische Information der Deutschen Nationalbibliothek: Die Deutsche Nationalbibliothek verzeichnet diese Publikation in der Deutschen Nationalbibliografie; detaillierte bibliografische Daten sind im Internet über http://dnb.d-nb.de abrufbar.

Alle in diesem Buch genannten Marken und Produktnamen unterliegen warenzeichen-, marken- oder patentrechtlichem Schutz bzw. sind Warenzeichen oder eingetragene Warenzeichen der jeweiligen Inhaber. Die Wiedergabe von Marken, Produktnamen, Gebrauchsnamen, Handelsnamen, Warenbezeichnungen u.s.w. in diesem Werk berechtigt auch ohne besondere Kennzeichnung nicht zu der Annahme, dass solche Namen im Sinne der Warenzeichen- und Markenschutzgesetzgebung als frei zu betrachten wären und daher von jedermann benutzt werden dürften.

Bibliographic information published by the Deutsche Nationalbibliothek: The Deutsche Nationalbibliothek lists this publication in the Deutsche Nationalbibliografie; detailed bibliographic data are available in the Internet at http://dnb.d-nb.de.

Any brand names and product names mentioned in this book are subject to trademark, brand or patent protection and are trademarks or registered trademarks of their respective holders. The use of brand names, product names, common names, trade names, product descriptions etc. even without a particular marking in this works is in no way to be construed to mean that such names may be regarded as unrestricted in respect of trademark and brand protection legislation and could thus be used by anyone.

Coverbild / Cover image: www.ingimage.com

Verlag / Publisher:
Südwestdeutscher Verlag für Hochschulschriften
ist ein Imprint der / is a trademark of
AV Akademikerverlag GmbH & Co. KG
Heinrich-Böcking-Str. 6-8, 66121 Saarbrücken, Deutschland / Germany
Email: info@svh-verlag.de

Herstellung: siehe letzte Seite /
Printed at: see last page
ISBN: 978-3-8381-3675-2

Zugl. / Approved by: Zürich, Universität Zürich, Dissertation, 2012

Copyright © 2013 AV Akademikerverlag GmbH & Co. KG
Alle Rechte vorbehalten. / All rights reserved. Saarbrücken 2013

Acknowledgments

Great thanks go to Prof. Dr. Lutz Jäncke as my supervisor during the three years of the doctoral thesis. I would like to thank him for his unceasing support on all levels, for the opportunity to creatively do research with at the same time receiving valuable mentoring in research questions. I always felt happy at the Division Neuropsychology.

Dr. Jürgen Hänggi was essential for my work. I very much enjoyed and appreciated co-working with him and I am very grateful for his support!

During the time at the Psychiatric University Hospital, I obtained valuable insights into clinical neuroscience. Special thanks go to Prof. Dr. Uwe Herwig and to Dr. Annette Brühl for their support and mentoring. I am very grateful for the allocated resources and the opportunity to include a patient sample in my studies.

I would like to thank Prof. Dr. Stephan Neuhauss for his support and for being part of the PhD steering committee.

Thanks go to all subjects who took part in magnetic resonance imaging for the present studies. Thanks to the complete Division Neuropsychology – I am glad to have been part of it. Thanks to Jutta Ernst who I met again coincidentally at the Burghölzli. Thanks to Marcel Herrmann who is a great friend. Thanks to my family for strong background support.

Table of contents

Abbreviations	I
List of figures	II
List of tables	III
Summary	IV
Zusammenfassung	V

1	**Background and aims**		1
1.1	Social anxiety disorder		1
1.2	Methodological introduction		1
	1.2.1	The brain's white matter and diffusion tensor imaging	1
	1.2.2	Automated parcellation of grey matter structures	5
1.3	Aims of the studies		6
2	**Empirical part**		7
2.1	Overview		7
2.2	Study 1		8
	Abstract		9
	Introduction		10
	Methods		11
	Results		16
	Discussion		20
	Conclusions		23
	Acknowledgments and funding		24
2.3	Study 2		25
	Abstract		26
	Introduction		27
	Methods		29
	Results		33
	Discussion		36
	Conclusions		40

	Acknowledgments and funding	41
2.4	Study 3	42
	Abstract	43
	Introduction	44
	Methods	46
	Results	49
	Discussion	52
	Conclusions	55
	Acknowledgments and funding	55
2.5	Author contributions	57
3	**General discussion**	**58**
3.1	Synopsis	58
3.2	Open questions	59
3.3	Limitations of the present studies	62
3.4	Concluding remarks	62
4	**References**	**64**

Appendix

Supplementary Material of Study 1	A
Supplementary Material of Study 2	E
Supplementary Material of Study 3	H

Abbreviations

ARES	Action Regulating Emotion Systems
ASI-3	Anxiety Sensitivity Index 3
BDI	Beck Depression Inventory
CSF	cerebrospinal fluid
DSM-IV	Diagnostic and Statistical Manual of Mental Disorders, 4th edition
DTI	diffusion tensor imaging
EPI	Eysenck Personality Inventory
EPI-L	Eysenck Personality Inventory ("lie" subscale)
FA	fractional anisotropy
FSL	FMRIB Software Library
FWE	family-wise error
FWHM	full width at half maximum
GM	grey matter
HC	healthy controls
IFOF	inferior fronto-occipital fasciculus
LSAS	Liebowitz Social Anxiety Scale
MNI	Montreal Neurological Institute
MRI	magnetic resonance imaging
OFC	orbitofrontal cortex
ROI	region of interest
SAD	social anxiety disorder
SAS	Soziale Angststörung
SD	standard deviation
SLF	superior longitudinal fasciculus
STAI	Spielberger State-Trait Anxiety Inventory
UF	uncinate fasciculus
WM	white matter

List of figures

1 Background and aims

Figure 1: The diffusion ellipsoid _____ 3
Figure 2: Examples of brain tissue sites and respective indexing on a
 qualitative scale of anisotropy _____ 4
Figure 3: Example of a tractography result _____ 5

2 Empirical part

Study 1

Figure 1: Group comparison of fractional anisotropy _____ 17
Figure 2: Voxel-wise correlation of fractional anisotropy with trait anxiety _____ 19
Figure 3: Region-of-interest analysis of the uncinate fasciculus _____ 20

Study 2

Figure 1: The uncinate fasciculus and the inferior fronto-occipital fasciculus _____ 28
Figure 2: Tracking procedure and examples _____ 33
Figure 3: Volume of the reconstructed fiber tracts in patients with
 social anxiety disorder and healthy subjects _____ 36

Study 3

Figure 1: Topography of the uncinate fasciculus and amygdala _____ 45
Figure 2: Triangular associations between trait anxiety, uncinate fasciculus
 volume, and amygdala volume _____ 52

List of tables

2 Empirical part

Study 1

Table 1: Demographic, psychometric and clinical characteristics of the sample _____ 12
Table 2: Group comparison of fractional anisotropy _____ 17
Table 3: Post-hoc analysis of diffusivity for the significant clusters obtained
from the group comparison of fractional anisotropy _____ 18

Study 2

Table 1: Demographic, psychometric and clinical measures _____ 34
Table 2: Tract-specific and global measures of interest _____ 35

Study 3

Table 1: Demographic and psychometric measures _____ 49
Table 2: Associations between white matter/grey matter volumes and trait anxiety _____ 51
Table 3: Associations between white matter and subcortical grey matter volumes _____ 51

Summary

As being a relatively frequent anxiety disorder, social anxiety disorder (SAD) increasingly attracts neuroscientists to characterize altered mechanisms in the brain that may promote or be a marker of the disorder. Whereas SAD-specific correlates of brain activation have been studied extensively, brain structure, especially white matter containing the large fiber bundles, remains poorly characterized with regard to altered mechanisms in SAD. The aim of the present thesis was to fill this gap and to complement previous studies. Three studies were conducted. The first two studies confirm a previously proposed role of the uncinate fasciculus (UF) for the pathophysiology of SAD. As a main outcome of the second study and as a new finding in the literature, UF volume was reduced in patients with SAD. The UF is a fiber bundle that links the orbitofrontal cortex with anterior temporal areas such as the temporal pole and the amygdala. The amygdala has previously been shown to be hyperactive in SAD and is implicated in the pathophysiology of anxiety disorders. Emotion regulation mechanisms, possibly centered on the orbitofrontal cortex as a higher-order evaluative and control region, are deficient in anxiety disorders. The identified alterations of the UF in the present studies, thus, fit into the concept of SAD pathophysiology. The third study was conducted to further establish volumetric measures of the UF. In an independent, non-clinical sample, UF volume was negatively correlated with trait anxiety and with amygdala volume. This underlines the significance of UF volumetric measures already pointed out in the second study. In addition, it suggests a role of the UF for more general mechanisms of anxiety. Thus, the UF might also represent a white matter structure basically linked to affect and stress – factors that are of high relevance across a range of psychiatric disorders.

Zusammenfassung

Als eine recht häufige Angststörung ist die Soziale Angststörung (SAS) mehr und mehr im Interesse der Neurowissenschaftler mit dem Ziel veränderte Mechanismen im Gehirn als Marker der Krankheit darzustellen. Neuronale Korrelate der SAS auf hirnfunktioneller Ebene sind mit grosser Studienanzahl untersucht worden. Hingegen ist die Hirn*struktur*, vor allem die aus grossen Faserbündeln bestehende weisse Substanz, hinsichtlich der Pathophysiologie der SAS noch wenig erforscht. Ziel der vorliegenden Dissertation war es diese Lücke zu schliessen und frühere Studien zu ergänzen. Es wurden drei Studien durchgeführt. Die ersten zwei Studien bekräftigen die durch eine frühere Studie postulierte Rolle des Fasciculus uncinatus (uncinate fasciculus, UF) für die Pathophysiologie der SAS. Das Hauptergebnis der zweiten Studie war ein verringertes Volumen des UF bei Patienten mit SAS, was zugleich ein neues Forschungsergebnis darstellt. Der UF ist ein Faserbündel, das den Orbitofrontalkortex mit vorderen temporalen Regionen wie dem Temporalpol und der Amygdala verbindet. Erhöhte Aktivität der Amygdala bei Patienten mit SAS wurde in früheren Studien gezeigt. Die Amygdala spielt daher eine wichtige Rolle in der Pathophysiologie der SAS. Kontroll- und Emotionsregulationsprozesse sind mit Aktivität im Orbitofrontalkortex verbunden und bei Angststörungen beeinträchtigt. Daher passen die in den vorliegenden Studien herausgestellten Veränderungen im UF zu Annahmen über die Pathophysiologie der SAS. Das Ziel der dritten Studie war unter anderem, die Bedeutung des UF-Volumens als Mass für Ängstlichkeitsprozesse zu unterstreichen. In einer unabhängigen Studienpopulation bestehend aus gesunden Versuchspersonen korrelierte das Volumen des UF negativ mit der Ängstlichkeit und mit dem Amygdala-Volumen. Dies bekräftigt die erstmals aus Studie 2 hervorgegangene Bedeutung des UF-Volumens. Ausserdem legt Studie 3 nahe, dass der UF eine generelle, nicht nur auf die SAS bezogene Rolle hinsichtlich Angst hat. In dieser Hinsicht könnte der UF sogar eine Struktur sein, die im Allgemeinen mit Emotionen und Stress zusammenhängt. Dadurch könnten Veränderungen im UF grundsätzlich mit der Vulnerabilität für die Entwicklung einer psychischen Störung in Verbindung stehen.

1 Background and aims

1.1 Social anxiety disorder

Anxiety disorders have a prevalence of about 15 % world-wide (Somers et al., 2006). Social anxiety disorder (SAD), also called social phobia, is one of the most common anxiety disorders, with an early age of onset and considerable heritability (Stein and Stein, 2008). The ȋgeneralizedȋ subtype of SAD is characterized by increased fear and avoidance of a range of social situations as stated in the *Diagnostic and Statistical Manual of Mental Disorders* (DSM-IV (American Psychiatric Association, 1994)), including

- being in focus of attention of others
- meeting previously unknown people and holding a conversation
- situations associated with (potential) evaluation by others.

Individuals suffering from SAD exhibit increased symptoms of arousal when being faced with such situations (e.g., sweating), typically have low self-esteem, and are at higher risk for developing major depressive disorder (Stein and Stein, 2008). In clinical neuroscience, SAD is of increasing interest. There is a considerable body of studies that examined SAD-specific alterations in regional brain activity and in neurotransmission profiles, however, brain structure remains under-investigated (Freitas-Ferrari et al., 2010). The identification of disorder-specific correlates in brain structure might contribute to a better understanding of SAD. In addition, it reflects the vision of being able by means of neuroimaging methods to complement diagnosis and treatment.

1.2 Methodological introduction [1]

1.2.1 The brain's white matter and diffusion tensor imaging

The brain is a complex network that is based on dynamic interactions between different regions in grey matter (GM), which comprises dense aggregates of neuronal cell bodies. Information is transferred from one neuron to another via an axon. Axons are covered with myelin, a substrate that is primarily composed of lipids and proteins (Edgar and Griffiths, 2009). Myelin serves as an electrical insulator between intra- and extra-cellular fluid, thus, it

---

[1] This chapter gives a short introduction to the applied methods in the present thesis. Automated parcellation of grey matter structures is briefly described in 1.2.2, as it is a minor part appearing in Study 3 (and is described there in detail).

facilitates electrical signal propagation along the axonal axis and is a molecular basis for efficient inter-regional communication (Edgar and Griffiths, 2009). Groups of highly aligned axons with similar start- or endpoints form a fiber bundle (or fiber tract). A fiber bundle reflects the structural connectivity between remote GM regions. Fibers and fiber bundles shape the brain's white matter (WM).

An approach to study structural connectivity is to focus on the specific anatomical features of WM. Magnetic resonance imaging (MRI) has the advantage of high spatial resolution. Diffusion Tensor Imaging (DTI) is an MRI method that is particularly sensitive for the delineation of WM structure. DTI is based on the diffusion properties of water molecules in brain tissue. Diffusion is an inherent thermophysical process that occurs differentially in different brain tissues (Basser and Özarslan, 2009): In GM and cerebrospinal fluid, little anatomical restriction is present for the movement of particles. Hence, diffusion is undirected, with no preferential spatial orientation. In WM, conversely, axonal membranes and myelin sheaths represent anatomical borders of diffusion. As a consequence, diffusion is restricted and mainly directed along the axonal orientation. In GM, diffusion is widely ìisotropî, whereas in WM, diffusion is more ìanisotropî.

To obtain measures of diffusivity including anisotropy magnitude, several steps from data acquisition to data modeling are applied. DTI stores diffusion intensity values for many separate spatial fragments, called voxels, in a three-dimensional picture of the brain. One voxel is typically 8 mm^3. During DTI, diffusion is measured for each voxel along several (about 30) different spatial directions. For each direction, one brain map is obtained where the diffusion magnitude along that particular direction is encoded in each voxel. Finally, information of diffusion along any single direction obtained from the respective maps is mathematically integrated into a complete diffusion profile for each voxel, reflected by a diffusion tensor (3×3 symmetric matrix):

$$D = \begin{pmatrix} D_{xx} & D_{xy} & D_{xz} \\ D_{xy} & D_{yy} & D_{yz} \\ D_{xz} & D_{yz} & D_{zz} \end{pmatrix}$$

The diagonalized form of the tensor contains three eigenvalues (λ_1, λ_2, λ_3):

$$D_d = \begin{pmatrix} \lambda_1 & 0 & 0 \\ 0 & \lambda_2 & 0 \\ 0 & 0 & \lambda_3 \end{pmatrix}$$

These eigenvalues are related to three eigenvectors (V_1, V_2, V_3). The three eigenvectors are orthogonal to each other. The principal (strongest) diffusion direction of a voxel is represented by V_1 with its diffusion magnitude λ_1. The eigenvalues and eigenvectors can be visualized by an ellipsoid, which represents the spatial extent of diffusion for a given voxel (Figure 1).

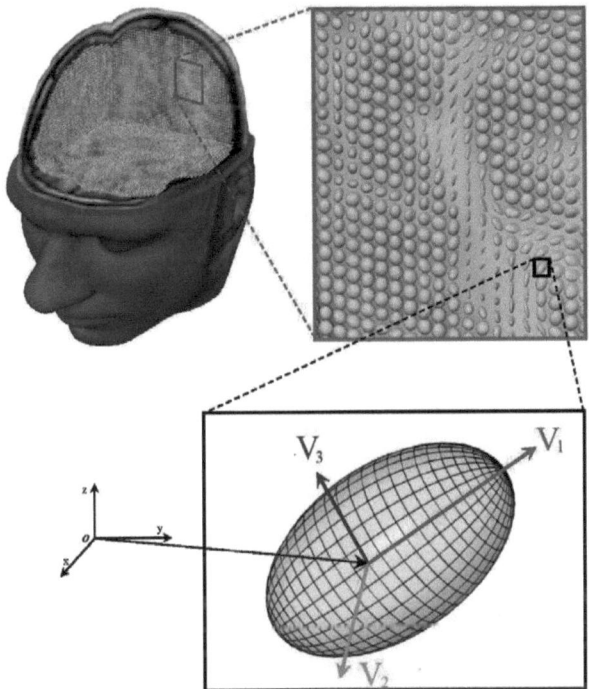

Figure 1: The diffusion ellipsoid
For each voxel in the brain volume, the diffusion tensor can be visualized by an ellipsoid. The ellipsoid is defined by three orthogonal axes (the eigenvectors V1, V2, V3) whose lengths correspond to the three eigenvalues, respectively ($\lambda_1, \lambda_2, \lambda_3$ with λ_1 having the largest value). The ellipsoid visualizes the diffusion space uniquely for each voxel. Diffusion preferentially occurs along the principal axis, defined by V1 and λ_1. [Figure adapted from Leemans (2006); permission obtained from the author]

The degree of anisotropy can be assessed according to a standardized measure, fractional anisotropy (FA) scaled between 0 and 1, given by

$$FA = \sqrt{\frac{1}{2}} \frac{\sqrt{(\lambda_1 - \lambda_2)^2 + (\lambda_2 - \lambda_3)^2 + (\lambda_1 - \lambda_3)^2}}{\sqrt{\lambda_1^2 + \lambda_2^2 + \lambda_3^2}}$$

In this formula, the differences in the numerator play a critical role: If all three diffusion magnitudes are similar ($\lambda_1 \approx \lambda_2 \approx \lambda_3$), FA gets near 0. The tensor visualization would then correspond to a sphere rather than an ellipsoid. In GM and cerebrospinal fluid, this would preferentially be the case. Conversely, it is apparent that the higher λ_1 is relative to λ_2 and λ_3, the higher gets FA. For instance, in the highly aligned WM of the corpus callosum, the magnitude of the principal diffusion direction is much higher as compared to the remaining two directions ($\lambda_1 \gg \lambda_2 \approx \lambda_3$), yielding a high FA value (of around 0.8). Correspondingly, the tensor would be cigar-shaped. Notably, as FA is determined by the three eigenvalues independent of the eigenvectors, it can be seen as a measure reflective of fiber *directedness*, but not of the fiber direction *itself*. As being a function of brain tissue, FA can be illustrated dimensionally on a scale (Figure 2) and be seen as indicator of different tissue classes (WM vs. GM) but also as sensitive of differential diffusion properties within a tissue class (highly vs. moderately aligned WM).

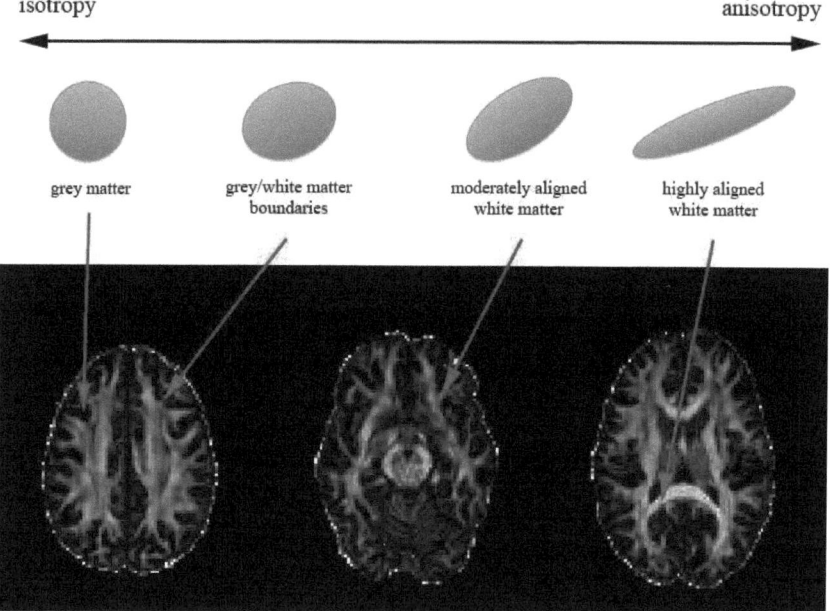

Figure 2: Examples of brain tissue sites and respective indexing on a qualitative scale of anisotropy
Simplified; respective ellipsoids are shown in blue.

Fiber tractography is a method to reconstruct and visualize fiber bundles based on preprocessed DTI images. "Tractography is the process of integrating voxel-wise fiber

1 Background and aims

orientations into a pathway that connects remote brain regions" (Behrens and Jbabdi, 2009). Commonly used tractography techniques determine a single orientation ("deterministic" tractography) of each voxel based on its principal eigenvector (V_1) and string together these orientations voxel per voxel. Tracts of interest are obtained by the definition of waypoints according to standard protocols based on specific anatomical landmarks that are penetrated by the tract (Wakana et al., 2007). Finally, trajectories ("fibers") are obtained that collectively form a representative of a fiber bundle (Figure 3). Tractography can be applied to create a mask of a tract of interest (e.g., for subsequent extraction of FA values within this mask) or to simply obtain the volume of the whole tract of interest.

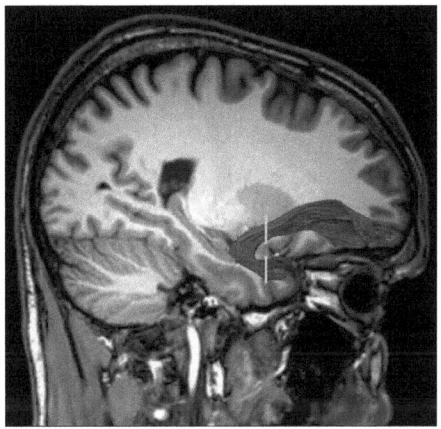

Figure 3: Example of a tractography result
The uncinate fasciculus, a fronto-temporal fiber bundle, is displayed in front of a T1-weighted high-resolution anatomical brain image. Two waypoints (displayed in yellow) are placed before the tractography process to define anatomical landmarks in the frontal and temporal lobe, respectively, to obtain the specific trajectories of the uncinate fasciculus.

FA and tract volume are quantitative measures of WM structural connectivity. Complementary to functional MRI studies, DTI has the potential to delineate anatomical underpinnings of inter-regional communication associated with personality traits (e.g., Xu and Potenza, 2012) and behavior (e.g., Schmithorst et al., 2011).

1.2.2 Automated parcellation of grey matter structures

With the development of the FreeSurfer software suite (surfer.nmr.mgh.harvard.edu), it has become possible to automatically analyze high-resolution T1-weighted structural MRI

1 Background and aims

images. One part of the FreeSurfer analysis pipeline is the volume-based stream "designed to preprocess MRI volumes and label subcortical tissue classes"[2]. With this method, each individual's subcortical structure is delineated based on different intensity values in the T1-image. After normalization into standard space, labeling of each structure is possible referring to a standard atlas. The several automatic processing stages are described in detail elsewhere (Fischl et al., 2002; Fischl et al., 2004).

1.3 Aims of the current thesis

In the current thesis, the first aim was to identify sites of altered WM connectivity in patients with SAD by means of assessing FA (Study 1). The second aim was to add to findings from Study 1 an alternative DTI processing approach focusing on tract volume (Study 2) to obtain a comprehensive view of possible mechanisms of SAD pathophysiology. The third aim was to validate the newly applied WM volumetric measure with regard to anxiety, testing evidence of a link to general anxiety-related mechanisms in a non-clinical, independent sample (Study 3).

---

[2] surfer.nmr.mgh.harvard.edu/fswiki/FreeSurferAnalysisPipelineOverview

2 Empirical part

2.1 Overview

Study 1

White matter alterations in social anxiety disorder

Published in Journal of Psychiatric Research 2011, 45(10): 1366-1372

Study 2

Evidence of frontotemporal structural hypoconnectivity in social anxiety disorder: a quantitative fiber tractography study

Published in Human Brain Mapping, in press

Study 3

Volumetric associations between uncinate fasciculus, amygdala, and trait anxiety

Published in BMC Neuroscience 2012, 13:4

2.2　Study 1

White matter alterations in social anxiety disorder

Published in Journal of Psychiatric Research 2011, 45(10): 1366-1372

Authors:

Volker Baur [1,2,*], Jürgen Hänggi [2], Michael Rufer [3], Aba Delsignore [3], Lutz Jäncke [2], Uwe Herwig [1], Annette Beatrix Brühl [1]

[1] Clinic for General and Social Psychiatry, Psychiatric University Hospital, University of Zurich, Switzerland

[2] Division Neuropsychology, Institute of Psychology, University of Zurich, Switzerland

[3] Department of Psychiatry and Psychotherapy, University Hospital Zurich, Switzerland

* Corresponding Author

Psychiatric University Hospital
Clinic for General and Social Psychiatry
Lenggstr. 31, CH-8032 Zurich, Switzerland

and (correspondence address):
Institute of Psychology, University of Zurich
Binzmühlestrasse 14/25, CH-8050 Zurich, Switzerland

Key words:

social anxiety disorder, diffusion tensor imaging, white matter connectivity, fractional anisotropy, uncinate fasciculus, trait anxiety

Abstract

White matter architecture in patients with social anxiety disorder (SAD) has rarely been investigated, but may yield insights with respect to altered structural brain connectivity. Initial evidence points to alterations in the uncinate fasciculus (UF). We applied diffusion tensor imaging in 25 patients with SAD and 25 matched healthy subjects. Whole-brain fractional anisotropy (FA) maps were used for group comparison and voxel-wise correlation with psychometric and clinical measures. Additionally, a region-of-interest analysis of the UF was performed. Patients with SAD had reduced FA compared to healthy subjects in or near the left UF and the left superior longitudinal fasciculus. There were no regions with increased FA in SAD. In the region-of-interest analysis, a negative correlation between FA and trait anxiety was identified in the left and right UF in patients, but not in healthy subjects. No correlations with social anxiety scores were observed. The present study partially confirms previous results pointing to frontal white matter alterations in or near the UF in patients with SAD. SAD-specific dimensional associations of FA with trait anxiety might reflect general pathological and/or compensatory mechanisms as a function of symptom severity in patients. Future studies should disentangle in which way the identified white matter alterations match functional alterations.

Introduction

Social anxiety disorder (SAD) is a common anxiety disorder (Jefferys, 1997) with a life-time prevalence of about 10 % (Kessler et al., 1994). It is characterized by exaggerated fear during the anticipation of or confrontation with evaluation by others. Since genetic factors are known to play a role in SAD (Stein and Stein, 2008; Mosing et al., 2009) and based on the commonly early age of onset (Stein and Stein, 2008), manifestations in the brain's structural architecture can be expected.

Examination of white matter (WM) may yield insights into structural connectivity and complement functional neuroimaging studies. Until now, little is known about WM architecture in SAD. First evidence points to alterations in the uncinate fasciculus (UF) (Phan et al., 2009), a fiber bundle connecting inferior frontal cortices including orbitofrontal cortex with the anterior temporal lobe and the amygdala (Ebeling and von Cramon, 1992; Ghashghaei et al., 2007; Petrides and Pandya, 2007). Since the amygdala mediates emotional arousal states (Davis and Whalen, 2001) and the orbitofrontal cortex is implicated in cognitive control of emotions (Ochsner and Gross, 2005), the UF may play a key role in the emergence or regulation of fear. Therefore, the UF represents a WM structure of interest for diffusion tensor imaging (DTI) studies related to fear and anxiety. Theories of SAD point to biases at different stages of cognitive-attentional processing (Hirsch and Clark, 2004), probably not mediated by the UF alone. Therefore, it can be expected that WM alterations are present in further brain areas mediating interpretative or associative processes.

DTI is a magnetic resonance based technique allowing the examination of WM structural properties, such as microstructural anisotropic diffusion reflected by fractional anisotropy (FA) (Basser and Pierpaoli, 1996). In the present study, we assessed FA throughout the whole brain in patients with SAD and in healthy control subjects. Additionally, we concentrated on FA within the UF in a region-of-interest approach. Given the results of related previous studies (Kim and Whalen, 2009; Phan et al., 2009), which for the first time investigated associations between FA and anxiety, we expected that reduced rather than increased FA would be a characteristic of abnormal WM architecture in anxious individuals. First, we aimed at identifying altered WM structures in patients with SAD compared to healthy subjects. Specifically, we hypothesized reduced FA in the UF. Further, we sought to determine the dimensional impact of trait anxiety, social anxiety and SAD duration on FA.

Methods

Subjects

Twenty-seven outpatients with current diagnosis of generalized SAD participated in this study. They were recruited from the outpatient clinic at the Department of Psychiatry and Psychotherapy of the University Hospital Zurich. Control subjects were recruited via direct address and email-advertisement. Due to severe artifacts in DTI images related to technical problems during data acquisition, two patients with SAD had to be excluded from further analyses. The remaining 25 patients with SAD were matched with 25 healthy subjects in terms of age and gender (see **table 1**). All subjects were right-handed according to a handedness questionnaire (Annett, 1970). Diagnosis of generalized SAD and comorbid Axis-I diagnoses were established using the Mini-International Neuropsychiatric Interview for DSM-IV (Sheehan et al., 1998, German version by Ackenheil et al., 1999). One patient met lifetime criteria for a depressive episode, but was not currently depressed; three patients met criteria for a current depressive episode in the course of a major depressive disorder, however, SAD was the primary diagnosis. One patient met criteria for prior alcohol dependency, currently remitted. There were no further psychiatric comorbidities. Altogether, nine patients were taking antidepressant medication due to depressive symptoms (selective serotonin reuptake inhibitors in five subjects, selective serotonin/norepinephrine reuptake inhibitors in two subjects, mirtazapine in one subject, and clomipramine/zolpidem in one subject). Two of the mentioned subjects were additionally taking lithium, one quetiapin. In all subjects, the dose of drugs had been stable for more than one month when participating in the study. Healthy control subjects were free of current or past psychiatric disorders (semi-structured diagnostic interview) and of medication (except oral contraceptives). Exclusion criteria in all subjects were neurological disorders, head trauma, pregnancy, excessive consumption of alcohol, cigarettes and caffeine, and contraindications against magnetic resonance imaging. All subjects provided written informed consent and were compensated for their participation. The study was approved by the local ethics committee.

General anxiety was measured with the trait version of the Spielberger State-Trait Anxiety Inventory (STAI (Spielberger et al., 1970), German version by Laux et al., 1981) in all participants. The degree of social anxiety was assessed in patients with the self-rating version of the Liebowitz Social Anxiety Scale (LSAS (Liebowitz, 1987), German version by Stangier and Heidenreich, 2005). The LSAS self-rating version has been shown to have high reliability (retest-reliability $r = 0.82$, Cronbach's $\alpha = 0.95$) and validity (correlation with the clinician-administered version $r \geq 0.78$) (Baker et al., 2002). Degree of depression in patients with SAD was assessed with Beck's Depression Inventory (BDI (Beck et al., 1961), German version by Hautzinger et al., 1995). Further patient-specific characteristics were collected in addition to the diagnostic interview: The duration of symptoms of SAD was assessed by retrospectively asking for the onset of symptoms (available in 20 patients). A positive family history of psychiatric disorders was present in 88.9 % of the patients (mostly SAD and depression, assessed in 18 patients).

Table 1: Demographic, psychometric and clinical characteristics of the sample

	SAD [a]	healthy subjects	test	T	df	p
age	32 ± 10.4 years (range: 19-53)	32 ± 10.1 years (range: 20-57)	t	0.24	48	0.82
gender	18 m, 7 f	18 m, 7 f				1.00
STAI-X2 [b]	[1]50 ± 11.2 (range: 31-76)	[2]33 ± 7.4 (range: 22-55)	t	6.38	41.6	< 0.001
LSAS [c]	66 ± 23.0 (range: 26-107)	-				
BDI [d]	15 ± 10.8 (range: 0-41)	-				
age of onset of SAD	15 ± 5.9 years (range: 6-30)	-				
duration of SAD	16 ± 10.6 years (range: 3-44)	-				

[a] SAD: patients with social anxiety disorder; [b] STAI-X2: Spielberger State-Trait Anxiety Inventory, trait section; [c] LSAS: Liebowitz Social Anxiety Scale; [d] BDI: Beck's Depression Inventory; [1]: corresponds to increased values (Laux et al., 1981), [2]: corresponds to values within the normal range (Laux et al., 1981)

DTI data acquisition

DTI scans were acquired on a 3.0 T whole-body scanner (GE Medical Systems, Milwaukee, USA) equipped with a standard 8-channel head coil. One diffusion-weighted spin-echo echo-planar imaging (EPI) scan was obtained from all participants. Slices were acquired sequentially in transversal orientation (matrix 256 x 256 pixels, 39 slices, slice thickness 3.2 mm, field of view (FOV) = 240 x 240 mm). Further imaging parameters were: echo time (TE)

= 87.8 ms, repetition time (TR) = 12000 ms. Diffusion sensitization was achieved with 2 balanced diffusion gradients centered on the 180° radio-frequency pulse. Diffusion was measured in 21 non-collinear directions with a b-value of $b = 1000$ s/mm^2. Five additional interleaved non-diffusion-weighted volumes ($b = 0$ s/mm^2) served as reference volumes. Scan time was about 6 minutes. In addition to DTI, T1- and T2-weighted images were consecutively acquired to exclude possible T1-/T2-sensitive tissue abnormalities.

Data preprocessing

We applied preprocessing procedures for DTI data as implemented in FMRIB Software Library (FSL, www.fmrib.ox.ac.uk/fsl) (Smith et al., 2004). Using FMRIB Diffusion Toolbox (FDT) (Behrens et al., 2003), FA maps as well as maps of primary (λ_1), secondary (λ_2) and tertiary (λ_3) eigenvalues were created. λ_1-maps were used for analysis of axial diffusivity. Radial diffusivity maps were calculated as $\lambda_{23} = (\lambda_2 + \lambda_3)/2$. Preprocessing comprised the following steps: 1) Eddy current and head movement correction were applied using FDT. 2) Individual binary brain masks were created on the non-diffusion weighted images using Brain Extraction Tool (Smith, 2002). 3) Tensors were fitted to the data using FDT. 4) Linear and non-linear normalization of the FA-maps into a standard stereo-tactic space (Montreal Neurological Institute, MNI; represented by the FMRIB58 FA template) were done with scripts (FLIRT for linear and FNIRT for non-linear normalization) implemented in FSL. 5) These transformations were then also applied to the λ_1- and λ_{23}-maps. 6) Images were resampled to a spatial resolution of 1 x 1 x 1 mm^3. 7) We visually inspected the quality of the resulting, normalized FA-, λ_1- and λ_{23}-maps.

Whole-brain analyses

Statistical analysis was done using statistical parametric mapping (SPM 5) software (www.fil.ion.ucl.ac.uk/spm). FA maps were smoothed with a Gaussian kernel of 6 mm FWHM (full width at half maximum) (e.g., see Park et al., 2004; Ha et al., 2009; Kim and Whalen, 2009) and thereafter thresholded considering only voxels with FA values greater than 10 % of global mean FA for statistical analysis. Applying general linear models to the FA maps, the significance of the differences between the two groups as well as of the correlations with psychometric and clinical measures was calculated by means of voxel-wise analysis of covariance. To examine effects of smoothing, we additionally applied smoothing with a Gaussian kernel of 3 mm FWHM and computed the voxel-based FA group comparison.

Differences between the two groups with respect to local FA were assessed using a correction for multiple comparisons of the statistical extent threshold combined with a non-stationarity smoothness correction (Worsley et al., 1999; Hayasaka and Nichols, 2004; Hayasaka et al., 2004). We used a voxel-wise threshold of $p < 0.00001$ uncorrected and a cluster-extent family-wise error (FWE) correction with $p < 0.05$ corrected for multiple comparisons (resulting in a cluster threshold of $k \geq 48$ voxels) for FA maps. Since FA is influenced by the ratio of axial to radial diffusivity, we examined post-hoc whether changes in FA in the respective cluster reflected changes in either axial or radial diffusivity or both. For each significant cluster in the group comparison of FA, we extracted post-hoc individual mean values of FA, axial and radial diffusivity from unsmoothed maps using MarsBaR toolbox of SPM 5 (Brett et al., 2002). These values were used to show group-specific magnitude values and to determine effect sizes (Cohen's *d*) of group differences. To assess medication-related effects on FA in patients with SAD, we divided the SAD group into two subgroups, medicated ($n = 9$) and medication-free ($n = 16$) patients. Age and trait anxiety were regressed out in these analyses to account for differences in these variables between the two SAD subgroups ($p = 0.11$ and $p < 0.05$, respectively). On the one hand, we examined post-hoc mean FA of the significant clusters obtained from the pooled FA group comparison (SAD vs. healthy subjects) using multivariate and univariate analysis of covariance with status of medication as between-subjects factor. On the other hand, we examined FA exploratory across the whole brain (medicated vs. medication-free patients) using the same statistical approach as for the pooled FA group comparison stated above.

In addition to the FA group comparison (SAD vs. healthy subjects), we correlated STAI and LSAS scores as well as SAD duration with local FA (Pearson correlation) in the whole brain. Here, we used a voxel-wise height threshold of $p < 0.001$ combined with a cluster-extent FWE correction for multiple comparisons ($p < 0.05$, corresponding to $k \geq 552$ for STAI, $k \geq 571$ for LSAS, and $k \geq 583$ for SAD duration). As in the FA group comparison, clusters from the correlation analyses were corrected for non-stationarity smoothness. We extracted mean FA, mean axial and mean radial diffusivity for each significant cluster from unsmoothed maps using MarsBaR toolbox of SPM 5 (Brett et al., 2002), which were then correlated post-hoc with the respective measure to obtain cluster-based correlation strength. Although trait anxiety was higher in patients with SAD compared to healthy subjects (mean ± standard deviation: 50 ± 11.2 vs. 33 ± 7.4, $p < 0.001$), which was to be expected, we assessed the voxel-wise correlation with STAI scores across all 50 subjects (25 patients and 25 healthy subjects). Because concerns may be raised on performing a correlation with STAI *across* both groups whereas a group difference with strong effect exists in this measure, we performed two post-hoc examinations. First, we correlated cluster-wise mean FA with trait anxiety separately for both groups to determine correlation strength *within* each of the two groups. Second, we computed step-wise linear regression for each identified cluster using "group" and "STAI" as independent variables and mean FA as dependent variable to investigate the way these factors distinctively explain variance of FA. In addition, we performed voxel-wise correlation with trait anxiety separately for both groups with the same statistical threshold as stated above. Correlations with LSAS and duration of SAD symptoms were restricted to the patient group.

With respect to demographic, psychometric, and global anatomical measures, we used independent two-tailed *t*-tests with a threshold of $p < 0.05$ uncorrected. WM regions obtained from group statistical analyses in SPM were overlaid on the group mean FA map in MNI space and their anatomical identification was based on the 'JHU White Matter Tractography Atlas' (Hua et al., 2008) and the 'Juelich Histological Atlas' (Eickhoff et al., 2005) implemented in FSL. To assure validity of our data, we inspected magnitude values of FA of each identified cluster separately for both groups with regard to plausibility for WM tissue (see also **table 2**).

Region-of-interest analysis

Due to previous findings in the UF (Kim and Whalen, 2009; Phan et al., 2009), we selected the left and right UF as regions of interest (ROI) based on probability maps from the 'JHU White Matter Tractography Atlas' (Hua et al., 2008). Probability was thresholded at 30 % both for the left and for the right UF ROI, excluding voxels with a low probability of being located within the UF. Resulting ROI sizes were 1074 mm^3 for the left UF and 488 mm^3 for the right UF ROI. In support of a valid analysis, we considered equal probabilities more important than equal sizes of left and right UF ROIs. For statistical analysis, voxel-wise FA mean values were extracted from individual, normalized and unsmoothed FA maps for both ROIs using MarsBaR toolbox of SPM 5 (Brett et al., 2002). Subsequently, these values were inspected for plausibility to be located in WM for each subject. To control for global mean FA, the ratio of within-UF mean FA to whole-brain mean FA (in the following referred to as "normalized" FA) was used for group comparison (independent two-tailed *t*-tests, $p < 0.05$) and correlation with trait anxiety (Pearson correlation and Spearman correlation). In addition to FA, axial and radial diffusivity were analyzed for both ROIs using the respective diffusivity maps.

Results

Psychometric, clinical and global anatomical measures
Psychometric, clinical and SAD-specific measures, including age of onset and duration of SAD symptoms, are summarized in **table 1**. Cross-correlations of STAI, LSAS and BDI are shown in **supplementary table S1** [1]. Patients had lower global mean FA (across all voxels of the brain) than healthy subjects (0.257 ± 0.007 vs. 0.262 ± 0.008, $p < 0.05$).

Local group differences of fractional anisotropy
Patients with SAD had reduced FA compared to healthy subjects in two regions ($p < 0.05$ cluster-extent FWE-corrected, see **table 2**): one located in the frontal lobe WM near the UF/inferior fronto-occipital fasciculus (see **figure 1A**), and the other located in the temporal lobe WM near the superior longitudinal fasciculus (see **figure 1B**). Conversely, there were no regions with increased FA in patients with SAD compared to healthy subjects at the same statistical threshold and even at a more liberal threshold of voxel-wise $p < 0.001$ uncorrected combined with a cluster-extent correction of $p < 0.05$. In both clusters with reduced FA in SAD, radial diffusivity was higher in SAD compared to healthy subjects, whereas in the temporal cluster, axial diffusivity was higher in healthy subjects (see **table 3**). There were no correlations of cluster mean FA with psychometric and clinical measures in the SAD or healthy subjects group. Of note, similar clusters of reduced FA in SAD were identified in the analysis based on 3-mm smoothed maps, with two additional clusters (see **supplementary figure S2** [3]).

Post-hoc analysis of the two significant clusters within the SAD group yielded no differences in FA between medicated and medication-free patients ($F = 0.735$, $p = 0.49$, see **supplementary table S3** [3] and **supplementary figure S4** [3] for details). Furthermore, there were no FA differences in the whole-brain exploratory approach between the SAD subgroups, even at a more liberal threshold of $p < 0.001$ uncorrected combined with $p < 0.05$ cluster extent FWE-corrected.

---

[3] see appendix, supplementary material of Study 1

2 Empirical part: Study 1

Table 2: Group comparison of fractional anisotropy (patients with social anxiety disorder < healthy subjects, $p < 0.00001$ voxel-wise uncorrected, $p < 0.05$ cluster-extent FWE-corrected)

cluster	size (mm³)	MNI peak coordinates			T (max)	FA (mean ± SD)		d
		x	y	z		SAD	HC	
UF/IFOF (left)	208	-24	38	-8	6.70	0.21 ± 0.03	0.26 ± 0.04	1.77
SLF (left)	185	-32	-31	17	6.70	0.38 ± 0.03	0.43 ± 0.03	1.56

UF: uncinate fasciculus, IFOF: inferior fronto-occipital fasciculus, SLF: superior longitudinal fasciculus; SAD: patients with social anxiety disorder; HC: healthy subjects

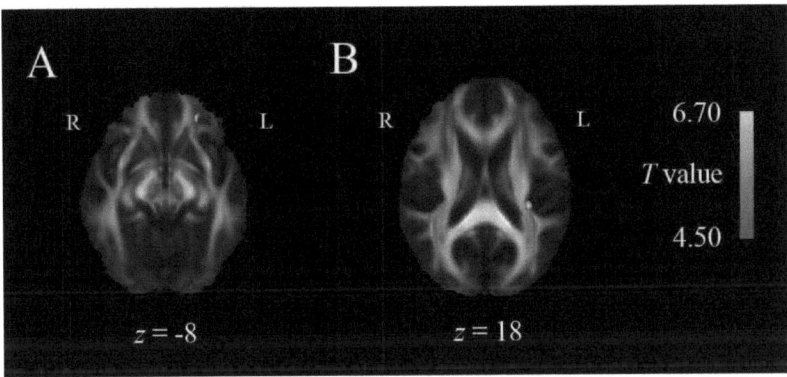

Figure 1: Group comparison of fractional anisotropy
Reduced fractional anisotropy (FA) in patients with social anxiety disorder compared to healthy subjects ($p < .05$ cluster-extent corrected) **(A)** in left prefrontal white matter near the uncinate fasciculus/inferior fronto-occipital fasciculus, **(B)** in left temporal white matter near the superior longitudinal fasciculus; clusters projected onto the group mean FA image in MNI space. Note: Clusters shown here are uncorrected for non-stationarity smoothness due to technical reasons. L: left, R: right

Table 3: Post-hoc analysis of diffusivity (* 1000 mm²/s) for the significant clusters obtained from the group comparison of fractional anisotropy

	cluster	eigenvalues (mean ± SD)		T	d
		SAD	HC		
axial diffusivity	UF/IFOF (left)	1.09 ± 0.07	1.07 ± 0.07	0.88	0.25
	SLF (left)	1.09 ± 0.05	1.12 ± 0.06	2.37	0.68
radial diffusivity	UF/IFOF (left)	0.81 ± 0.06	0.72 ± 0.06	4.97	1.43
	SLF (left)	0.60 ± 0.04	0.56 ± 0.03	4.59	1.39

UF: uncinate fasciculus, IFOF: inferior fronto-occipital fasciculus, SLF: superior longitudinal fasciculus; SAD: patients with social anxiety disorder; HC: healthy subjects; *T*-values (uncorrected) refer to two-tailed *t*-tests contrasting respective mean diffusivity values of patients with SAD and healthy subjects

Voxel-wise correlation of fractional anisotropy with trait anxiety, social anxiety and duration of social anxiety disorder symptoms

Voxel-wise correlation of FA with trait anxiety across all subjects yielded a negative correlation in two clusters (see **figure 2**), with the more lateral one ($r = -0.62$) encompassing regions of the left UF, and the more medial one ($r = -0.58$) encompassing subcortical structures as centromedial/extended amygdala and ventral striatum (see **supplementary table S5** [2]). Post-hoc analyses for these clusters are shown in **supplementary table S6** [4] and **supplementary figure S7** [4]. These analyses in summary suggested that cluster-based FA is modulated by trait anxiety rather than group, being more evident for the lateral cluster compared to the medial one. Voxel-wise correlation of FA with trait anxiety separately for SAD and healthy subjects yielded no significant clusters, as did voxel-wise correlation of FA with social anxiety in the SAD group. No significant correlation of duration of SAD symptoms with trait anxiety was observed ($r = 0.16$, $p = 0.51$). A negative correlation of FA with SAD duration was identified in WM near the right inferior temporal gyrus/fusiform gyrus ($r = -0.64$) (see **supplementary table S5** [4]). Cluster-based correlations with axial/radial diffusivity for all identified clusters are shown in **supplementary table S5** [4].

---

[4] see appendix, supplementary material of Study 1

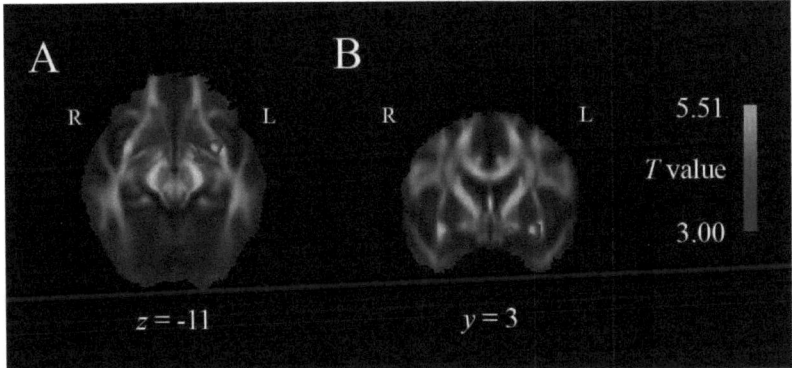

Figure 2: Voxel-wise correlation of fractional anisotropy with trait anxiety
Negative correlation of fractional anisotropy (FA) with trait anxiety across all subjects ($n = 50$, $p < 0.05$ cluster-extent corrected) in two clusters covering parts of the left uncinate fasciculus and connected subcortical structures, **(A)** axial view, **(B)** coronal view; clusters projected onto the group mean FA image in MNI space. Note: Clusters shown here are uncorrected for non-stationarity smoothness due to technical reasons. L: left, R: right

Region-of-interest analysis of the uncinate fasciculus
Absolute and normalized FA of the left and right UF was not different between groups (each $p \geq 0.12$). However, normalized FA correlated negatively with trait anxiety in patients with SAD, both in the left ($r = -0.41$, $p < 0.05$, see **figure 3**, Spearman $\rho = -0.41$, $p < 0.05$) and in the right UF ($r = -0.40$, $p = 0.05$, Spearman $\rho = -0.41$, $p < 0.05$), but not in healthy subjects ($p = 0.51$ and $p = 0.81$, respectively). Axial and radial diffusivity within the left and right UF were not different between groups (each $p \geq 0.23$). Axial diffusivity was negatively correlated with trait anxiety in the left UF across all subjects ($r = -0.30$, $p < 0.05$).

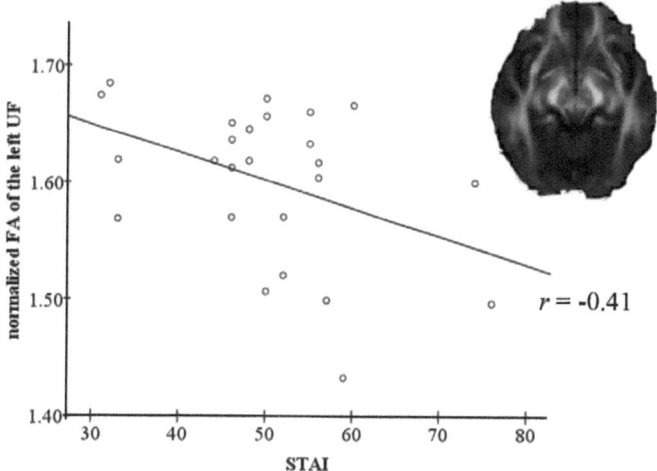

Figure 3: Region-of-interest analysis of the uncinate fasciculus
Plot of the negative correlation between normalized mean fractional anisotropy within the left uncinate fasciculus (region-of-interest analysis, cluster visualized in the upper right corner, cluster size 1074 mm^3) and trait anxiety in patients with social anxiety disorder (n = 25). FA: fractional anisotropy, UF: uncinate fasciculus, STAI: Spielberger State-Trait Anxiety Inventory (trait section)

Discussion

Our aim was to identify WM regions that may play a role in the pathophysiology of SAD with a particular interest in the UF due to previous evidence (Kim and Whalen, 2009; Phan et al., 2009). The main finding was reduced FA in patients with SAD compared to healthy subjects in a frontal region near the left UF and in a temporal region near the left superior longitudinal fasciculus. There were no regions in which FA was higher in SAD compared to healthy subjects. According to the ROI analysis, we identified a negative correlation of FA with trait anxiety both in the left and right UF in patients with SAD, but not in healthy subjects.

In the present study, reduced FA was in two regions categorically associated with the diagnosis of SAD and in other regions dimensionally with the magnitude of psychometric and clinical measures. In general, the exact meaning of FA measures at the anatomical level is still under debate (Kubicki, 2010) and it may not be justified to simply conclude from reduced FA to deficient or abnormal signal propagation along fiber bundles, which is a functional perspective (see also Imfeld et al., 2009; Jancke et al., 2009; Oechslin et al., 2009). Reduced FA reflects either reduced axial diffusivity related to compromised axonal structure, or increased radial diffusivity, or both (Mori and Zhang, 2006). We found two regions in which FA was reduced in patients with SAD compared to healthy subjects (see **figure 1**). In both regions, radial diffusivity was increased in SAD (see **table 3**). However, multiple scenarios such as, for instance, demyelination may lead to increased radial diffusivity (Song et al., 2002). The first cluster with reduced FA in SAD was located to the UF/inferior fronto-occipital fasciculus, partially confirming the initial finding by Phan and colleagues (Phan et al., 2009). There, reduced FA in SAD was identified focally in the right UF. Compared to that finding, the cluster of the present study differed with respect to hemisphere (left), location (more anterior), cluster-based magnitude of FA (lower) and effect size of the group difference (larger). Inspection of exact location (see **figure 1A**) and FA values (around 0.2, see **table 2**) suggests that this WM cluster is at the border to grey matter of the left orbitofrontal cortex and not directly located within the main trajectory of the UF. Hemisphere differences between our findings and those of Phan and colleagues point to the need for further DTI studies examining this aspect. Two recent functional imaging studies suggest that the left orbitofrontal cortex is involved in altered functional connectivity patterns with the amygdala in SAD (Liao et al., 2010; Hahn et al., 2011). Notably, FA of the frontal part of the left UF has been shown to be associated with the 5-HTTLPR polymorphism related to vulnerability for anxiety and depression (Pacheco et al., 2009). Besides frontal WM, our results suggest that an area in or near the superior longitudinal fasciculus in the temporal lobe WM may be involved in the pathophysiology of SAD. This cluster was located near the posterior insula (see **figure 1B**). Our finding of reduced global mean FA in SAD could point to even more regions in which FA is reduced and which were not detected by the present approach. Possibly, differences in such regions are of slight magnitude, and future studies that include larger sample sizes of patients with SAD may extend our findings by revealing additional WM regions with reduced FA. Overall, differences of the results compared to Phan and colleagues may be due to the different samples (with slight composition differences in age, sex and race), different number of diffusion directions, differences in smoothing (we used smoothed data), and different strategies to control for false-positives (we used an explicit correction for multiple comparisons over the cluster-extent at $p < 0.05$). Of note, sample sizes and spatial resolution were comparable in both studies.

In the voxel-wise correlation of FA with trait anxiety across all subjects, we identified two clusters being located side by side with a negative correlation covering parts of the left UF, extended amygdala (Heimer, 2003) and basal ganglia (see **figure 2**). We are aware that there may be conceptual issues concerning this correlation analysis, since there was a significant difference with a high effect between SAD and healthy subjects in trait anxiety, whereas the voxel-wise correlation was performed across all subjects (SAD and healthy subjects). Post-hoc analyses (see **supplementary material S6/S7** [5]) suggested that – at least for the lateral cluster located in the UF – the group difference in trait anxiety did not bias the correlation. Yet, we explicitly prompt caution in interpreting the results of this correlation analysis with regard to dimensional associations with general anxiety, especially for the medial cluster. It cannot be completely ruled out that disorder-related effects other than general anxiety have biased the results, since trait anxiety and group affiliation are highly correlated factors. In addition, we did not identify similar clusters in the voxel-wise correlation separately for the two groups. Notably, however, the mentioned clusters encompass posterior UF and connected subcortical structures, which replicates a recent study demonstrating in healthy subjects a negative correlation between FA and trait anxiety in very similar areas (Kim and Whalen, 2009). Furthermore, structures as the extended amygdala and ventral striatum are implicated in models of anxiety-related processes (LeDoux and Gorman, 2001). They may be considered part of cortico-striato-thalamo-cortical loops (Alexander et al., 1986) which have been implicated in the pathophysiology of a number of neuropsychiatric disorders (Heimer, 2003). Projections from the centromedial amygdala to the ventral striatum are considered important in the shift from passive-avoiding behavior towards active-coping reactions in response to fear-inducing stimuli (LeDoux and Gorman, 2001).

The ROI analysis in the bilateral UF according to our a-priori hypothesis based on previous findings of Phan and colleagues (Phan et al., 2009) revealed no significant group differences within the left and right UF, which was not according to our hypothesis. This was possibly due to the relatively large extent of the anatomically defined UF ROIs, if group differences were solely present in a small region within the ROIs. Although there were no group differences, FA correlated negatively with trait anxiety both in the left (see **figure 3**) and right UF in the SAD group. In healthy subjects, no such correlation could be identified. This might reflect that structural alterations are dimensionally associated with pathological mechanisms or increased compensatory efforts due to severe anxiety, which is present in patients but not in healthy subjects. Moreover, we found a negative correlation of axial diffusivity with trait anxiety in the left UF across all subjects. As the UF interconnects the amygdala and orbitofrontal cortex (Ghashghaei et al., 2007), it may mediate inhibitory control over the

---

[5] see appendix, supplementary material of Study 1

amygdala by frontal cortices, which may counteract pathological anxiety states (Bishop, 2007, 2009; Etkin et al., 2009). Albeit small, the identified correlation suggests that there might be a link between the integrity of UF fibers and the capacity of regulatory influence on anxiety. However, more research is still necessary to confirm this hypothesis.

Where may the identified WM alterations in SAD stem from? Because SAD has a considerable genetic component (Stein and Stein, 2008; Mosing et al., 2009) and an early age of onset (Stein and Stein, 2008), mechanisms promoting the disorder may be manifested in the brain's WM architecture, which might be reflected by altered FA. Obviously, it cannot be stated by the present study whether the identified changes of FA in patients with SAD stem from changes at an early developmental stage or from changes occurring in the course of the disorder. The latter may be particularly possible given the mean duration of SAD of 16 (± 10.6) years in the examined sample of patients. In this context, it also cannot be stated whether the identified WM alterations are specific to SAD. More unspecific experiences like frequent exposure to stress, which is present in any anxiety disorder, may also have lead to the observed group differences. DTI studies comparing different states of anxiety/anxiety disorders could help clarify this issue.

Limitations

The impact of medication on FA is still unknown. Several studies reported no association between psychotropic medication and FA (Yurgelun-Todd et al., 2007; McIntosh et al., 2008; Wang et al., 2008b; Wang et al., 2008a; Sussmann et al., 2009). Similarly, our analyses suggest that medication did not have any impact on FA in our sample. However, we are careful to generalize this finding to other sorts of medication and other psychiatric disorders. Another limitation is unequal statistical power (due to heterogeneous sample sizes) in the correlation analyses for trait anxiety, social anxiety and duration of SAD symptoms. This constrains the direct comparison of the resulting statistical maps to a certain extent. We suggest considering the results from the correlation analyses with caution.

Conclusions

Taken together, our results partially confirm and extend the initial study by Phan and colleagues (Phan et al., 2009). Evidence of frontal WM alterations in or near the UF in SAD converges, whereas there may be further WM structures being involved in the pathophysiology of SAD. More research on functional correlates of reduced FA has to be done in future studies, particularly to clarify the role of the UF and related structures in fear, anxiety, general emotion processing, and emotion regulation.

Acknowledgments and funding

We thank Beat Werner, University Children's Hospital Zurich, for technical assistance with DTI data acquisition. Funding for this study was provided by a grant of the Swiss National Foundation to UH.

2.3 Study 2

Evidence of frontotemporal structural hypoconnectivity in social anxiety disorder: a quantitative fiber tractography study

Published in Human Brain Mapping, in press

Authors:

Volker Baur [1,*], Annette Beatrix Brühl [2], Uwe Herwig [2], Tanja Eberle [1], Michael Rufer [3], Aba Delsignore [3], Lutz Jäncke [1,4,5], Jürgen Hänggi [1]

[1] Division Neuropsychology, Institute of Psychology, University of Zurich, Switzerland

[2] Clinic for General and Social Psychiatry, Psychiatric University Hospital, University of Zurich, Switzerland

[3] Department of Psychiatry and Psychotherapy, University Hospital Zurich, University of Zurich, Switzerland

[4] Center for Integrative Human Physiology, University of Zurich, Switzerland

[5] International Normal Aging and Plasticity Imaging Center (INAPIC), University of Zurich, Switzerland

* Corresponding Author

Institute of Psychology, University of Zurich
Binzmühlestrasse 14/25, CH-8050 Zurich, Switzerland

Key words:
social anxiety disorder, diffusion tensor imaging, quantitative fiber tractography, white matter connectivity, uncinate fasciculus, volume

Abstract

Investigation of the brain's white matter fiber tracts in social anxiety disorder (SAD) may provide insight into the underlying pathophysiology. Because models of pathological anxiety posit altered fronto-limbic interactions, the uncinate fasciculus (UF) connecting (orbito-) frontal and temporal areas including the amygdala is of particular interest. Microstructural alterations in parts of the UF have been reported previously, whereas examination of the UF as discrete fiber tract with regard to more large-scale properties is still lacking. Diffusion tensor imaging was applied in 25 patients with generalized SAD and 25 healthy control subjects matched by age and gender. By means of fiber tractography, the UF was reconstructed for each participant. The inferior fronto-occipital fasciculus (IFOF), originating from the frontal cortex similarly to the UF, was additionally included as control tract. Volume and fractional anisotropy (FA) were compared between the groups for both tracts. Volume of left and right UF was reduced in patients with SAD, reaching statistical significance for the left UF. Bilateral IFOF volume was not different between groups. A similar pattern was observed for FA. Reduced volume of the left UF in SAD fits well into pathophysiological models of anxiety, as it suggests deficient structural connectivity between higher-level control areas in the orbitofrontal cortex and more basal limbic areas like the amygdala. The results point to a specific role of the left UF with regard to altered white matter volume in SAD. However, results should be replicated and functional correlates of altered UF volume be determined in future studies.

Introduction

Social anxiety disorder (SAD) involves intense fear and avoidance of social situations, e.g. being in focus of attention of others, with about one of ten in the general population meeting criteria for SAD during life-time (Kessler et al., 1994). Hyperactive limbic and paralimbic areas like the amygdala and insula are central to the pathophysiology across a range of anxiety disorders including SAD, post-traumatic stress disorder and specific phobia (Etkin and Wager, 2007). The amygdala has a pivotal role in salience processing with a particular relatedness to social cues (Adolphs, 2003a) and is, therefore, of special importance regarding social anxiety (Cannistraro and Rauch, 2003). Beyond hyperactivity of basal affective systems, alterations in SAD have also been shown in frontal cortical activity relevant for cognitive control processes (Goldin et al., 2009), for example in the orbitofrontal cortex (OFC). Models of pathological anxiety posit impaired functional dialogue between basal affective and higher-order control systems (Cannistraro and Rauch, 2003; Akirav and Maroun, 2007; Bishop, 2007), which may underlie the observed functional alterations in limbic and frontal areas in SAD.

The specific features of white matter (WM) architecture in SAD, especially with regard to important fiber bundles, have not been studied extensively so far. However, a more detailed examination of these anatomical features could help understand the neural underpinnings of SAD. The uncinate fasciculus (UF) connects the OFC with limbic/paralimbic regions including amygdala and the anterior temporal lobe (Ebeling and von Cramon, 1992; Petrides and Pandya, 2007). Hereby, the UF may provide the anatomical connection underlying the functional dialogue between higher-order control and basal affective brain systems. Given also the particular relatedness of amygdala and OFC function to social processes (Adolphs, 2003b) and, thus, social anxiety, the UF is a main fiber tract of interest for the investigation of SAD pathophysiology. Initial evidence points to local microstructural alterations in parts of the right UF in patients with SAD (Phan et al., 2009), underlining our focus on the UF in the present investigation. Examination of the UF as discrete fiber tract with regard to more large-scale properties such as volume is still lacking in the literature.

There are different approaches to study the WM architecture in SAD. Previously (Baur et al., 2011), we focused on fractional anisotropy (FA), a measure modulated by fiber directedness. Assessing FA, it is possible to estimate fiber disorganization and demyelination in SAD on a micro-/mesoscopic level. Here, we used an alternative approach in the same subjects: Diffusion tensor imaging (DTI) quantitative fiber tractography allows for the reconstruction and quantification of whole fiber bundles within the individual brain's WM (Mori and van Zijl, 2002). Because the prominent, well-known WM fiber tracts facilitate functional interactions between distant brain regions, they may also support integration of different modalities (e.g., sensory, cognitive, and emotional processes). For instance, the UF might facilitate cognitive-emotional interactions by enabling communication between the OFC and the amygdala, that is, between higher-level control and basal affective areas. Implications from UF tractography may therefore refer to a more large-scale level regarding the pathophysiology of SAD. Here, we focused on large-scale aspects of the UF and hypothesized reduced UF volume in SAD, corresponding to deficient fronto-limbic interactions. To examine whether possible volume differences in the UF could be attributed to unspecific or global effects, we included the inferior fronto-occipital fasciculus (IFOF) interconnecting frontal and occipital lobes as control tract for fiber tracking and statistical analysis. The UF and the IFOF share a common trajectory in the frontal lobe (Catani et al., 2002) (see also **figure 1**) and separate in more posterior regions to follow their own trajectories terminating in the temporal and occipital lobe, respectively. For the IFOF, we did not expect any group differences since there is no direct evidence that this tract is involved in the pathophysiology of SAD. In addition to volume, we assessed mean FA of each reconstructed tract. Analogously, we expected reduced FA of the UF, but not of the IFOF, in SAD.

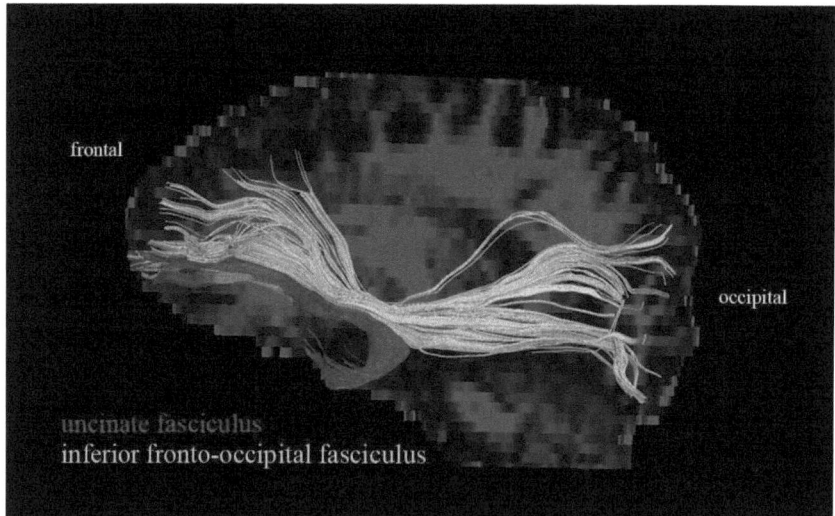

Figure 1: The uncinate fasciculus (red) and the inferior fronto-occipital fasciculus (yellow)
Example of their relative locations (shown for one subject, lateral view onto the left hemisphere)

Methods

Subjects

For the study, 27 outpatients with current diagnosis of generalized SAD were recruited from the outpatient clinic at the Department of Psychiatry and Psychotherapy of the University Hospital Zurich, Switzerland. Due to severe artifacts in DTI images, two patients with SAD had to be excluded from further analyses. In addition to the remaining 25 patients, 25 healthy control subjects were recruited via direct address and email-advertisement. Patient and control groups were matched by age and gender (see **table 1**). All subjects were consistently right-handed according to the procedure provided by Annett (Annett, 1970). Diagnosis of generalized SAD and current and previous comorbid Axis-I diagnoses were established in the patients group using the Mini-International Neuropsychiatric Interview for DSM-IV (Sheehan et al., 1998) (German version (Ackenheil et al., 1999)). SAD was the primary diagnosis in all patients, five fulfilled criteria for comorbidities (former depressive episode (remitted) in one patient, current depressive episode/major depressive disorder in three patients, alcohol dependency (remitted) in one patient). Nine patients were taking antidepressant medication due to reactive depressive symptoms (selective serotonin reuptake inhibitors in five patients, selective serotonin/norepinephrine reuptake inhibitors in two patients, mirtazapine in one patient, and clomipramine/zolpidem in another patient). Two of these patients were additionally taking lithium, one quetiapin. Dose of drugs had been stable for more than one month in all nine patients when participating in the study. Healthy control subjects were free of current or past psychiatric disorders and of medication (except for oral contraceptives in females), as determined in a semi-structured clinical interview according to DSM-IV. Neurological disorders, head trauma, pregnancy, excessive consumption of drugs (alcohol, nicotine, caffeine) and further contraindications against magnetic resonance imaging served as exclusion criteria for the study. For assessment of general anxiety, all participants completed the trait version of the Spielberger State-Trait Anxiety Inventory (STAI (Spielberger et al., 1970), German version (Laux et al., 1981)). Social anxiety was assessed in patients with the self-rating version of the Liebowitz Social Anxiety Scale (LSAS (Liebowitz, 1987), German version (Stangier and Heidenreich, 2005)). Degree of depression in patients with SAD was assessed with Beck's Depression Inventory (BDI (Beck et al., 1961), German version (Hautzinger et al., 1995)). In addition, patients were asked to retrospectively state the onset of their symptoms. The majority of patients (88.9 %) reported a positive family history of psychiatric disorders (mostly SAD and depression, information available in only 18 patients). After complete description of the study to the subjects, written informed consent was obtained. The study was approved by the local ethics committee.

DTI data acquisition

DTI scans were acquired on a 3.0 T whole-body scanner (GE Medical Systems, Milwaukee, USA) equipped with a standard 8-channel head coil. One diffusion-weighted spin-echo echo-planar imaging (EPI) scan was obtained from all participants. Slices were acquired sequentially in transversal orientation (matrix 256 x 256 pixels, 39 slices, slice thickness 3.2 mm, field of view (FOV) = 240 x 240 mm^2, in-plane spatial resolution 0.94 x 0.94 mm^2). Further imaging parameters were: echo time (TE) = 87.8 ms, repetition time (TR) = 12000 ms. Diffusion sensitization was achieved with 2 balanced diffusion gradients centered on the 180° radio-frequency pulse. Diffusion was measured in 21 non-collinear directions with a *b*-value of $b = 1000$ s/mm^2. Five additional interleaved non-diffusion-weighted volumes ($b = 0$ s/mm^2) served as reference volumes. Scan time was about 6 minutes. In addition to DTI, T1- and T2-weighted images were acquired to exclude possible T1-/T2-sensitive abnormalities.

Data preprocessing and fiber tractography

Preprocessing was done with FMRIB Software Library (FSL) Version 4.1.6 (Smith et al., 2004) (www.fmrib.ox.ac.uk/fsl) and comprised the following steps: 1) segregation of brain tissue from non-brain tissue using the Brain Extraction Tool (Smith, 2002); 2) Eddy current and head movement correction using EDDYCORRECT from FMRIB's Diffusion Toolbox (Smith et al., 2004); 3) rotation of the gradients according to the corrected parameters from step 2); 4) local fitting of diffusion tensors and construction of individual FA maps using DTIFIT from FMRIB's Diffusion Toolbox (Smith et al., 2004).

For fiber tracking, Diffusion Toolkit 0.5 and TrackVis 0.5.1 were used (Wang et al., 2007) (www.trackvis.org). The preprocessed data from FSL were further processed with Diffusion Toolkit. For each subject, the diffusion tensors were estimated according to the corrected gradients. Deterministic fiber tracking was performed with the "brute-force" approach (Huang et al., 2004), an automatic procedure commonly used to reconstruct fibers across the whole WM by tracking fibers from each voxel in the brain. The fiber assignment continuous tracking (FACT) algorithm (Mori et al., 1999) was used. Accordingly, fibers were reconstructed by TrackVis along the principal eigenvector of each voxel's diffusion tensor. Tracking termination criteria were angle > 45° and FA < 0.2 (Mori and van Zijl, 2002) (individual FA map derived from FSL's DTIFIT was used as mask image in Diffusion

Toolkit). Fiber tracking was performed successively in each subject's native space. Color-coded FA maps derived from the principal eigenvector of the diffusion tensor in each voxel were used for region-of-interest (ROI) drawing in TrackVis. ROIs were drawn large-sized to include the entirety of the tract of interest and avoid false-negative fibers (Yasmin et al., 2009) (see also **figure 2**). All fiber tracts were obtained through a two-ROI approach (seed ROI and target ROI) with logical AND concatenation (Catani et al., 2002; Wakana et al., 2007) of the two ROIs, such that only fibers that passed both ROIs were included in the reconstructed tract. Obviously spurious fibers were removed from the fiber tract by using an additional avoidance ROI (logical NOT operation) (Wakana et al., 2007). For the UF, both the seed and the target ROI was placed in the same coronal slice where the anterior-posterior fibers (coded in green) of the frontal and the temporal lobe were visible at the most posterior point (see **figure 2A** for illustration of the ROI placement and tractography examples for the UF, see also (Wakana et al., 2007)). For the IFOF, the seed ROI was placed in the occipital lobe according to Wakana and colleagues (Wakana et al., 2007). The target ROI was placed at the densest portion of the fiber bundle projecting anteriorly (coded in green, anterior floor of the external capsule (Catani et al., 2002)), typically located in the coronal slice that dissects the middle of the corpus callosum body (see **figure 2B** for illustration of the ROI placement and tractography examples for the IFOF). Each tract was reconstructed in both hemispheres, and tracking was randomly performed either first in the left or in the right hemisphere in each subject. After tractography, each individual tract was visually inspected for plausibility with regard to its structure based on general anatomical knowledge and previously published tractography studies (Catani et al., 2002; Mori et al., 2002; Wakana et al., 2007). For each tract, any voxel touched by a fiber was counted by TrackVis. As such, volume values were obtained by accumulating all voxels belonging to the respective tract.

Tractography was performed by two investigators (VB and TE) blinded to group affiliation of the subjects. Both investigators did exactly the same steps (ROI placement etc.) of the tracking procedure as described above. Tracking results from the first investigator were used for statistical analysis. The second investigator reconstructed all tracts for 16 randomly chosen subjects (eight belonging to the patients group and eight to the control group). The values obtained for these tracts were used to determine inter-rater reliability.

Statistical analysis

For each subject and each reconstructed fiber tract (left UF, right UF, left IFOF, right IFOF), the following variables were extracted from TrackVis: volume (in ml), mean FA, mean fiber length (in mm), and fiber count (artificial unit for number of fibers). In addition, global (whole-brain) values according to the "brute-force" tracking approach delivered by TrackVis were obtained for each of these measures. Statistical analysis was done with IBM SPSS Statistics (Version 19, SPSS Inc, an IBM company, Armonk, NY, USA).

Demographic and psychometric group differences were examined using independent two-tailed t-tests ($p < 0.05$). Inter-rater reliability was assessed by calculating the intra-class correlation coefficient on absolute volume and absolute FA values for the tracts of 16 subjects measured by both raters. For examination of normal distribution, Kolmogorov-Smirnov test was used.

Group differences of the UF and IFOF with regard to volume and FA were assessed with independent t-tests contrasting the relative tract measures (tract volume divided by global WM volume; tract mean FA divided by global mean FA). Our focus was on UF volume, for which we had a directional hypothesis of reduced volume in patients with SAD compared to healthy subjects. Because we tested both left and right UF, cumulation of alpha error was controlled for by applying Bonferroni correction resulting in a corrected $\alpha = 0.025$ for the left and right UF, respectively. Significance tests related to the IFOF were performed thereafter under the hypothesis *not* to see group differences. Thus, Bonferroni correction was applied solely for tests related to the UF. Analysis of FA was done accordingly. To assess possible medication-related effects of volume and FA, post-hoc, univariate analysis of covariance (ANCOVA) was used with status of medication as between-subjects factor ($n = 9$ vs. $n = 16$) and with the respective global measure, trait anxiety (difference between SAD subgroups at $p < 0.05$), and age (difference between SAD subgroups at $p = 0.11$) as covariates of no interest. To assess possible comorbidity-related effects of volume and FA, post-hoc, ANCOVA comprised status of comorbidity as between-subjects factor ($n = 5$ vs. $n = 20$) with respective global measures as covariates of no interest. Further post-hoc analysis related to volume was done investigating relative mean fiber length and relative number of fibers (independent t-tests and Pearson correlation). Dimensional associations of relative tract volume as well as relative tract mean FA with trait anxiety (STAI) and social anxiety (LSAS) were examined with Pearson's correlation. p-values related to UF volume and FA were one-tailed (corrected for multiple comparisons), whereas all other p-values were two-tailed (uncorrected). Additionally, effect sizes (Cohen's d) were determined.

2 Empirical part: Study 2

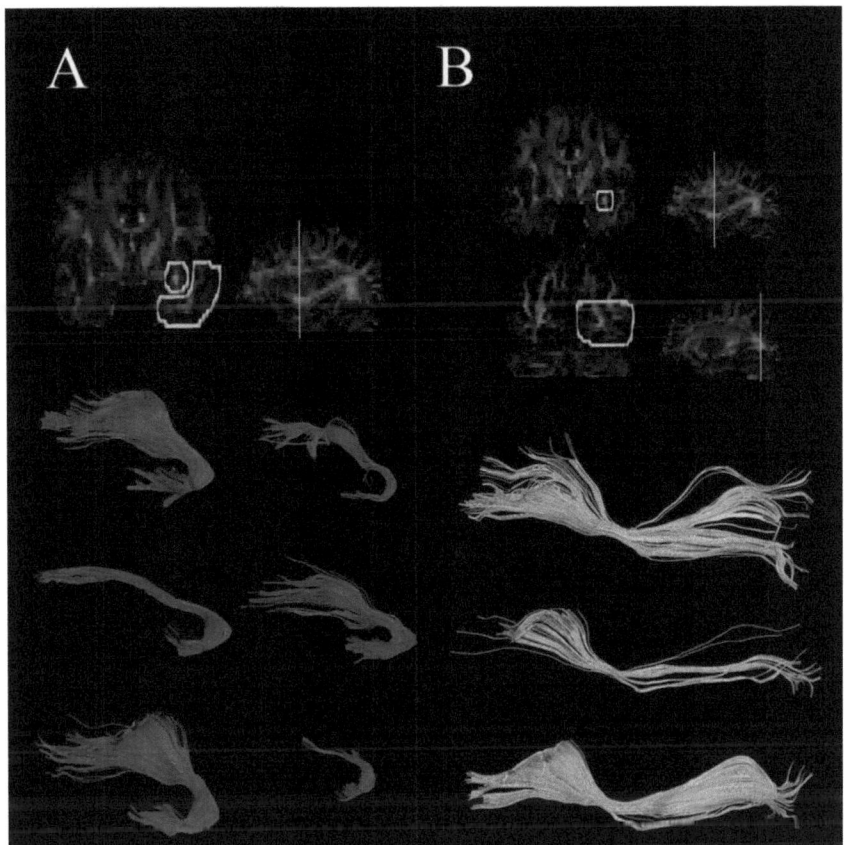

Figure 2: Tracking procedure and examples

A. Tracking of the uncinate fasciculus (UF): Placement of ROIs within one coronal slice (left) and localization of this slice in a lateral view (right) are shown. For illustration purposes, six examples of the left UF are shown in red.

B. Tracking of the inferior fronto-occipital fasciculus (IFOF): Placement of ROIs within two different coronal slices (left) and respective localization of these slices in a lateral view (right) are shown. For illustration purposes, three examples of the left IFOF are shown in yellow.

Results

General measures, inter-rater reliability and normal distribution

Demographic, psychometric and clinical measures are summarized in **table 1**. Intra-class correlation coefficients were > 0.92 for absolute volume and > 0.95 for absolute mean FA, indicating excellent reproducibility for the main measures of interest for each reconstructed fiber tract. Kolmogorov-Smirnov tests yielded normal distributions for each fiber tract, with regard to both relative volume (each $p > 0.57$) and relative FA (each $p > 0.90$).

Table 1: Demographic, psychometric and clinical measures

	SAD		HC		t^a	p
	mean	SD	mean	SD		
age (yrs)	31.6	10.4	32.3	10.1	0.24	0.82
STAI	50.2 [b]	11.1	33.2 [c]	7.4	6.38	< 0.001
LSAS	66.0	23.0				
BDI	15.2	11.0				
age of onset (yrs) [d]	15.1	6.0				
duration of symptoms (yrs) [d]	15.5	10.9				
	male	female	male	female		
sex	18	7	18	7		

[a] SAD vs. HC, according to an independent t-test; [b] corresponds to increased values (Laux et al., 1981); [c] corresponds to normal values (Laux et al., 1981); [d] available in 20 patients

SAD: patients with social anxiety disorder, HC: healthy controls, SD: standard deviation, STAI: Spielberger State-Trait Anxiety Inventory (trait version), LSAS: Liebowitz Social Anxiety Scale, BDI: Beck's Depression Inventory

Volume and fractional anisotropy associated with social anxiety disorder

Tractography was successful for all tracts and all subjects. Absolute volume and FA values for the reconstructed tracts and global WM are shown in **table 2**, separately for patients with SAD and for healthy subjects, including measures of the respective statistical comparisons.

As the main result, patients with SAD had a significantly lower relative volume of the left UF compared to healthy subjects ($p_{Bonferroni} = 0.024$). Volume of the right UF was also reduced in SAD, however, without reaching a statistical significance or trend ($p_{Bonferroni} = 0.088$). For the left and right IFOF and global WM, no significant differences were observed (see **table 2, figure 3**). There were no correlations of volume with psychometric and clinical measures (see **supplementary table S1** [6]). Post-hoc analysis related to left UF volume yielded no significant differences, for neither medication ($F = 2.34$, $p = 0.14$) nor comorbidity ($F = 0.65$, $p = 0.80$). Further post-hoc examination related to left UF volume (summarized in **supplementary table S2** [6]) yielded significantly reduced mean fiber length compared to healthy subjects, whereas fiber count was not significantly different between groups. However, variance in left UF volume was rather explained by fiber count (87 %) as compared to fiber length (61 %).

---

[6] see appendix, supplementary material of Study 2

Beyond examination of volume, patients with SAD had lower relative mean FA than healthy subjects of the left UF at a trend level ($p_{Bonferroni} = 0.0495$), whereas for the other tracts, no significant differences or trends were observed (see **table 2, supplementary figure S3** [7]). Global mean FA was significantly reduced in patients with SAD. A correlation of relative FA with trait anxiety was observed for the left UF and right IFOF in SAD, but not in healthy subjects (see **supplementary table S1, supplementary figure S4** [7]). There were no correlations with social anxiety. Post-hoc analysis related to left UF FA yielded neither medication- ($F = 1.82, p = 0.19$) nor comorbidity-related ($F = 1.38, p = 0.25$) effects.

Table 2: Tract-specific and global measures of interest

			SAD		HC		t^a	p	d^b
			mean	SD	mean	SD			
volume [ml]	UF	left	4.51	2.39	5.77	2.03	2.03	0.024 [d]	0.61
		right	5.01	1.69	5.60	1.45	1.37	0.088 [d]	0.40
		p^c	0.25		0.51				
	IFOF	left	8.54	2.27	9.49	3.63	0.91	0.37	0.29
		right	8.41	2.37	8.75	2.23	0.22	0.82	0.06
		p^c	0.77		0.19				
	global WM		730.2	81.7	739.7	65.0	0.46	0.65	0.13
mean FA	UF	left	0.445	0.027	0.464	0.023	1.68	0.0495 [d]	0.49
		right	0.452	0.025	0.455	0.023	0.81	0.21 [d]	0.23
		p^c	0.17		< 0.05				
	IFOF	left	0.524	0.022	0.532	0.022	0.34	0.73	0.10
		right	0.517	0.028	0.524	0.025	0.32	0.75	0.09
		p^c	0.07		< 0.05				
	global WM		0.487	0.013	0.496	0.013	2.39	0.021	0.69

[a] SAD vs. HC, according to an independent t-test contrasting relative values for UF and IFOF (local tract value divided by global WM value); [b] Cohen's d (effect size); [c] left vs. right, according to a paired t-test; [d] one-tailed, corresponding to a corrected $\alpha = 0.025$ according to Bonferroni

UF: uncinate fasciculus, IFOF: inferior fronto-occipital fasciculus, FA: fractional anisotropy, SAD: patients with social anxiety disorder, HC: healthy controls, WM: white matter

---

[7] see appendix, supplementary material of Study 2

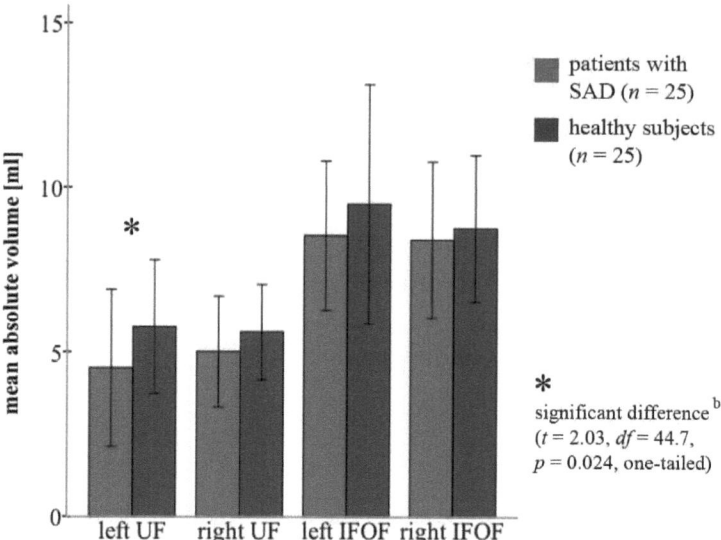

Figure 3: Volume of the reconstructed fiber tracts in patients with social anxiety disorder and healthy subjects

Mean (bars) and standard deviation (error bars) of absolute volume values are shown for each tract and group. SAD: social anxiety disorder; UF: uncinate fasciculus; IFOF: inferior fronto-occipital fasciculus; [b] according to a t-test contrasting relative volume values (ratio of tract volume and global WM volume)

Discussion

We used quantitative fiber tractography to investigate differences of volume and FA between patients with SAD and healthy subjects in two fiber tracts originating in the frontal cortex. Our main finding was reduced volume of the left UF in SAD, which accords with our hypothesis. There were no statistically significant group differences for the right UF and bilaterally for the IFOF. WM fiber tracts in the brain provide the anatomical basis for direct functional interactions between distant brain regions, facilitating also integrative processes. The UF connects frontal cortices including OFC with anterior temporal areas and with the amygdala (Ebeling and von Cramon, 1992; Petrides and Pandya, 2007) and may therefore facilitate fronto-limbic interactions. In anxiety disorders, activation of limbic/paralimbic areas as the amygdala and insula is increased (Cannistraro and Rauch, 2003; Etkin and Wager, 2007), and models of exaggerated anxiety additionally implicate deficient prefrontal control mechanisms in anxious subjects (Bishop, 2007, 2009; Freitas-Ferrari et al., 2010). Being faced with social situations, patients with SAD exhibit strong emotional responding, which may correspond to exaggerated activity of the amygdala and diminished prefrontal cognitive control. Cognitive-emotional integration such as emotion regulation strategies may be crucial for dealing with stressors and tends to be disturbed in anxiety disorders like SAD (Turk et al., 2005; Salters-Pedneault et al., 2006). If control/evaluative systems (OFC) and salience/affective systems (amygdala, insula) lack exchange of information, limbic circuits will possibly develop "a life of its own" leading to exaggerated arousal states (Freitas-Ferrari et al., 2010) which are frequent in SAD. Since the UF may be the main tract facilitating direct functional interactions between the OFC and the amygdala (Ebeling and von Cramon, 1992; Petrides and Pandya, 2007), which may involve top-down inhibition (Ghashghaei and Barbas, 2002), reduced volume of the left UF fits well into models of SAD pathophysiology. Reduced volume in the left UF suggests structural hypoconnectivity between grey matter areas in the frontal and anterior temporal lobe, which may yield implications for neuronal communication encompassed by these areas. For instance, functional connectivity between the OFC and the amygdala has been shown to be crucial for cognitive reappraisal of negative stimuli (Kanske et al., 2011) and capable of decreasing negative affect (Banks et al., 2007). Both structures are implicated in evaluation of significance of stimuli, fear extinction, and decision making based on their functional dialogue (Dolan, 2007). Furthermore, they may underlie approach and avoidance behavior (Aupperle and Paulus, 2010). Recently, effective connectivity between the OFC and the amygdala has been shown to be increased in SAD bidirectionally (Liao et al., 2010). In the light of our present results, this would point to compensatory mechanisms on a functional level due to a structural "deficit". Evidence of compensatory mechanisms in SAD comes from another recent study assessing a relationship between reduced grey matter volume in temporal cortices and increased functional connectivity with these areas (Liao et al., 2011). Besides hyperactivity of the amygdala (Cannistraro and Rauch, 2003; Etkin and Wager, 2007), under-recruitment of the OFC associated with social anxiety has been shown in functional neuroimaging studies (Tillfors et al., 2001; Bruhl et al., 2011; Zhou et al., 2011). The fact that we found no correlations with measures reflecting social anxiety challenges the notion whether reduced UF volume is of specific relevance for SAD. Rather, the present results might point to a pathophysiologic characteristic underlying abnormal anxiety or mood regulation in general. This is supported by studies that have identified WM alterations in the UF in mood disorders and schizophrenia (McIntosh et al., 2008; Kawashima et al., 2009), suggesting in turn a role of the UF in processes modulated by common risk factors of several psychiatric disorders, such as early-life stress. Comparative studies across different psychiatric disorders would be necessary to address this issue.

On a morphometric level, we consider three possibilities that may explain the reported volume difference of the left UF between patients with SAD and healthy subjects. First, fiber loss and thus reduced fiber density may lead to reduced volume in SAD. This might also explain the finding of reduced FA within the left UF. Second, there are more large-scale interconnected and/or a higher number of involved grey matter areas associated with the UF in healthy subjects, reflected by thinner and/or shorter UF tracts in SAD. Third, it is a mixture of both. Indeed, our post-hoc analysis yielded reduced mean fiber length of the left UF in SAD and 87 % explained variance of left UF volume by fiber count. This indicates that both the length and pure presence of fibers have contributed to the effect of reduced volume.

Analysis of FA yielded a similar pattern as for volume: Patients with SAD had reduced mean FA of the left UF (statistical trend with moderate effect size), whereas for the other reconstructed tracts, there were no differences compared to healthy subjects. This points to micro-/mesoscopic alterations in WM along or in distinct portions of the left UF, for example changes in fiber orientation and/or organization, and partially confirms a previous study in which reduced FA in SAD was identified in a part of the right UF (Phan et al., 2009). Further findings of the present study were reduced global mean FA in SAD and a negative correlation of left UF FA and trait anxiety in patients with SAD, but not in healthy subjects. A detailed discussion of FA alterations in this sample of patients with SAD and a more detailed comparison to findings by Phan and colleagues can be found in our previous report (Baur et al., 2011).

Reduced global mean FA in SAD is in contrast to global WM volume, for which we did not find group differences. Possibly, FA is reduced in a spatially more diffuse manner, whereas WM volume is reduced more specifically (e.g., in the left UF). It was not within the scope of the present study to assess associations between FA and volume for individual fiber tracts or globally for the brain's WM. Future studies have to address this question explicitly. Although our results indicate that WM alterations in SAD relate to volume in addition to FA, it may be of note that the reported volumes are related to the applied tractography preprocessing steps (see also Wakana et al., 2007). We used an FA threshold of 0.2 as recommended and used in other tractography studies as well (Mori and van Zijl, 2002; Rodrigo et al., 2007; Wakana et al., 2007). Thus, volume values refer to white matter in which FA is greater than 0.2. The two ROIs used for tractography of the UF were located in the same coronal slice slightly anterior to the temporal horn (see **figure 2A**). Hence, there may be more tolerance for variance in fiber length in the frontal part as compared to the temporal part of the UF. Since mean UF fiber length was reduced in SAD besides evidence of reduced FA in SAD in orbitofrontal/frontopolar WM according to previous reports (Phan et al., 2009; Baur et al., 2011), we cannot completely rule out that the volume effect in the left UF is partially influenced by cases in the SAD group with some voxels in the left frontopolar area having FA smaller than 0.2 but still belonging to UF WM. This would, however, in turn suggest impaired integrity of fibers in the left UF, probably involving a lack of frontopolar links to more posterior and temporal parts (like the insula, amygdala and temporal pole) in SAD (see **supplementary animated picture S5** [8]). Three issues support the present approach: First, setting the threshold more liberally (e.g., 0.1) would have resulted in including more spurious and false-positive fibers and, thus, hampered the tractography process. Second, having identified globally reduced FA in SAD but not globally reduced volume suggests that reduced volume of the UF is not just a "covered" effect of reduced FA. Third, fiber count heavily contributed to the finding of reduced left UF volume.

It may be worth to point out that within our sample there were no downward statistical outliers for left UF volume in the SAD group. Although there was a high variation related to volume and shape in the reconstructed tracts, our UF volume mean and standard deviation values of the healthy subjects are in line with those reported by Hasan and colleagues (Hasan et al., 2009). We included each tract as it was initially reconstructed after the tracking procedure (see methods), for the reason that this might reveal potential features of underlying SAD pathophysiology. However, our finding needs to be replicated. The choice of including the IFOF as control tract stems from the fact that, just like the UF, it originates in the frontopolar cortex. As in the frontal lobe the IFOF's trajectory runs at close quarters to the UF (see **figure 1**), inspection of IFOF trajectory served as a means to rule out possible biases related to data acquisition (e.g., frontal signal drop-outs). With regard to contents, both tracts mediate intra-hemispheric communication with frontal cortices, which stand for complex human processes like worrying and thus may be one source of anxiety disorders (Berkowitz et al., 2007).

Limitations

Nine of 25 patients in the present study sample were taking medication, which represents a certain limitation. We decided to include the medicated patients for three reasons. First, other DTI studies in anxiety disorders used a similar strategy (Ha et al., 2009; Phan et al., 2009),

---

[8] The animated picture can be downloaded at onlinelibrary.wiley.com/doi/10.1002/hbm.21447/suppinfo.

which makes the present study comparable to those studies. Second, long-term effects of antidepressant medication on brain structure due to plasticity, however, may rather lead to adaptation towards healthy subjects' brain morphology (Castren, 2005), not biasing the volume effect found in the present study. Third, medicated patients had significantly higher anxiety levels than those without medication. Since the volume effect of the left UF disappears when contrasting medication-free patients ($n = 16$) vs. healthy subjects, it is likely that inclusion of the medicated patients favored the detection of volume alterations in SAD that may be related to elevated anxiety. This reflects an area of conflict between the missing of real effects due to exclusion and the detection of biased effects due to inclusion of medicated patients. The latter seems unlikely according to the applied post-hoc examination focusing on the left UF revealing neither statistical significance nor trends of possible medication-related effects. However, the potential influence of medication on WM structure in general should be paid attention on by researchers in clinical neuroscience, even more unless there are studies that explicitly investigate dose-dependent impact of antidepressants on WM. Because, to the best of our knowledge, this is the first tractography study in anxiety disorders, we believe that the present approach may be perceived as justified and be of importance for future studies applying this method to patients with anxiety disorders.

A methodological limitation relates to DTI data acquisition: Here, in-plane resolution was high (< 1 mm^2), whereas resolution along the z-axis was relatively low (> 3 mm). Because isometric voxel sizes are generally recommended for DTI tractography studies, this prompts additional caution in interpreting the results of UF fiber tracking.

Implications of the present results for WM structure in SAD can only bear on, and thus are limited to, the actual reconstructed tracts, namely the UF and IFOF. Examination of further prominent fiber tracts with regard to volume would have been beyond the scope of our present study. *Global* WM volume not being significantly reduced in SAD (see **table 2**), however, is in favor of the view that volume reduction in SAD may indeed be specific to the left UF. Yet, further well-known fiber tracts may be included in future tractography studies in SAD.

Conclusions

To the best of our knowledge, this is the first study reporting on fiber tract volume alterations in SAD. Quantitative fiber tractography may be a useful tool to investigate anatomical WM connectivity within well-known fiber tracts in SAD. We were able to show smaller volume and FA values in patients with SAD for the left UF, but not for the right UF and the IFOF. This suggests particular importance of fronto*temporal* WM presence concerning the pathophysiology of SAD, possibly because of the facilitation of cognitive-emotional interactions between the OFC and the amygdala through the UF. Three topics of significance beyond SAD may emerge from the present results and guide future studies: 1) identification of functional correlates of UF volume alterations, 2) characterization of the relationship between FA and WM volume, and 3) comparative investigation of the UF's role in/across different psychiatric disorders.

Acknowledgments and funding

We thank Beat Werner, University Children's Hospital Zurich, for technical assistance with DTI data acquisition. Funding for this study was provided by a grant of the University of Zurich to LJ and a grant of the Swiss National Foundation (120518) to UH.

2.4 Study 3
Volumetric associations between uncinate fasciculus, amygdala, and trait anxiety

Published in BMC Neuroscience 2012, 13:4

Authors:

Volker Baur [1,*], Jürgen Hänggi [1], Lutz Jäncke [1,2,3]

[1] Division Neuropsychology, Institute of Psychology, University of Zurich, Switzerland
[2] Center for Integrative Human Physiology, University of Zurich, Switzerland
[3] International Normal Aging and Plasticity Imaging Center (INAPIC), University of Zurich, Switzerland

[*] Corresponding author

Institute of Psychology, University of Zurich
Binzmühlestr. 14/25, CH-8050 Zurich, Switzerland

Key words:
trait anxiety, uncinate fasciculus, amygdala, hippocampus, volume, white matter, grey matter, tractography, diffusion tensor imaging, subcortical segmentation

Abstract

Recent investigations of white matter (WM) connectivity suggest an important role of the uncinate fasciculus (UF), connecting anterior temporal areas including the amygdala with prefrontal-/orbitofrontal cortices, for anxiety-related processes. Volume of the UF, however, has rarely been investigated, but may be an important measure of structural connectivity underlying limbic neuronal circuits associated with anxiety. Since UF volumetric measures are newly applied measures, it is necessary to cross-validate them using further neural and behavioral indicators of anxiety. In a group of 32 subjects not reporting any history of psychiatric disorders, we identified a negative correlation between left UF volume and trait anxiety, a finding that is in line with previous results. On the other hand, volume of the left amygdala, which is strongly connected with the UF, was positively correlated with trait anxiety. In addition, volumes of the left UF and left amygdala were inversely associated. The present study emphasizes the role of the left UF as candidate WM fiber bundle associated with anxiety-related processes and suggests that fiber bundle volume is a WM measure of particular interest. Moreover, these results substantiate the structural relatedness of UF and amygdala by a non-invasive imaging method. The UF-amygdala complex may be pivotal for the control of trait anxiety.

Introduction

A growing body of neuroimaging studies links white matter (WM) measures to anxiety-related psychological processes (Kim and Whalen, 2009; Phan et al., 2009; Baur et al., 2011; Hettema et al., 2012). These studies in summary point to the uncinate fasciculus (UF), a prominent fronto-temporal fiber tract known to innervate the amygdala (Thiebaut de Schotten et al., 2012; Klingler and Gloor, 1960; Ebeling and von Cramon, 1992; Petrides and Pandya, 2007). The amygdala has been shown to be part of a limbic network that is hyper-responsive in individuals with increased anxiety (Stein et al., 2007) and in patients with anxiety disorders (Etkin and Wager, 2007). In this network, the UF is a pivotal part providing a link from the amygdala to prefrontal/orbitofrontal cortical areas, and thus is involved in modulating anxiety (Kim et al., 2011). Therefore, the UF is of particular interest for investigating the relation between anxiety and WM morphometry.

Using fiber tractography, we previously showed that patients with social anxiety disorder compared to healthy subjects demonstrate reduced volume of the left UF, suggesting fronto-temporal structural hypoconnectivity (Baur et al., in press). Regarding the preliminary nature of this result, identification of correlates of UF volume in a completely independent study sample is needed to validate UF volumetric measure and to further characterize in what way left UF volume is associated with anxiety. Anxiety is complex, involving emotional, cognitive, motivational, and physiological components and is dimensional with high inter-individual variability (Laux et al., 1981). Personality traits are seen as intra-individual stable factors, which can be linked to brain structure (e.g., Omura et al., 2005; Wright et al., 2006). *Trait anxiety* is defined as a psychological construct including several components, which merge and result in feelings such as discomfort, nervousness, and unpleasantness. Trait anxiety is a fixed stage of anxiety existing for a relative long duration or is even stable over a longer time period. Strong trait anxiety is also seen as the propensity to become extra-anxious in the context of provoking stimuli (Laux et al., 1981). Here, we sought to explore the dimensional association of UF volume with trait anxiety as assessed by means of the widely used Spielberger State-Trait Anxiety Inventory (STAI) (Laux et al., 1981). In addition to WM, we also investigated grey matter (GM) volumes of subcortical structures. The amygdala-hippocampus complex has been implicated in fear- and anxiety-related behavior (Bannerman et al., 2004). Functional imaging studies have also demonstrated a link between individual differences in trait anxiety and amygdala activity by showing increased amygdala activity with increasing anxiety (Etkin et al., 2004; Stein et al., 2007; Carlson et al., 2011;

Sehlmeyer et al., 2011). Regarding volumetric associations with trait anxiety, two studies point to a prominent role of the left amygdala compared to the other subcortical structures (Spampinato et al., 2009; Blackmon et al., 2011). However, studies focusing on the specific anxiety-related neuroanatomical features of the amygdala and linked pathways are still rare. In addition, these studies do not consistently report anxiety-related volumetric associations for the amygdala (see also Kuhn et al., 2011). Thus, we focus – beside the UF – on the amygdala GM volume and its relatedness to anxiety as well as its relatedness to UF volume (for the topography of UF and amygdala, see **figure 1**).

In the present study, we investigated volumes of the UF, the amygdala, and the hippocampus in a non-clinical sample of 32 subjects with different levels of trait anxiety. For this study, we formulated the following hypotheses:

(i) UF volume should correlate negatively with trait anxiety (according to the previous finding of reduced left UF volume in social anxiety disorder (Baur et al., in press));

(ii) The amygdala volume should correlate positively or negatively with trait anxiety;

(iii) UF and amygdala volumes should be strongly intercorrelated either negatively or positively, depending also on the association between trait anxiety and amygdala volume;

Since the anterior hippocampus has been associated with anxiety and this hippocampal part is also tightly linked to the amygdala (amygdala-hippocampus complex) (Bannerman et al., 2004), we also anticipated a correlation between trait anxiety and hippocampus volume.

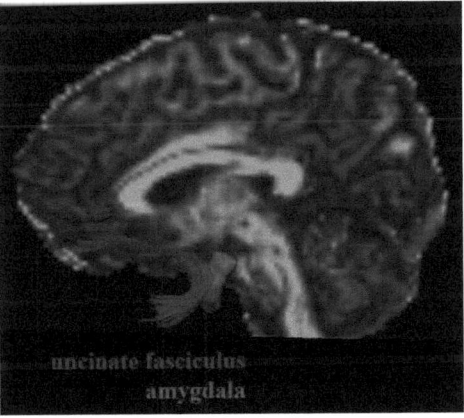

Figure 1: Topography of the uncinate fasciculus and amygdala
Example of their relative locations, shown for one subject in native space, lateral view onto the left hemisphere (right hemisphere on background)

Methods

Subjects

We called attention to the study by advertisement on mailing lists and notice boards in the University Zurich buildings. In the first part of this project, 218 subjects took part at an online-screening test where they completed the STAI (trait section) (Laux et al., 1981) and specified socio-demographic characteristics. In addition, they were asked with respect to exclusion criteria, which were general contraindications against magnetic resonance imaging (MRI), consummation of drugs, excessive consummation of alcohol and nicotine, medication affecting the central nervous system, known history of neurologic or psychiatric disorders, pregnancy, and age over 40. Subjects were selected based on their anxiety levels assessed in the online-screening in order to achieve an equal distribution of trait anxiety in the MRI study group. Finally, 35 healthy subjects (18 female, 17 male) were asked and were willing to participate in the MRI study. Absence of exclusion criteria was confirmed for each subject prior to scanning. None of the subjects reported any current or past neurologic and psychiatric disorders. Three subjects reported a history of psychotherapeutic treatment. None of the subjects reported any current medication. One participant reported a history of antidepressant medication. This participant was excluded from the study. To determine discriminant validity of associations with trait anxiety, further affect-related traits were assessed in addition to the STAI. Before scanning, subjects completed the Anxiety Sensitivity Index questionnaire (ASI-3) (Reiss et al., 1986), Beck Depression Inventory (BDI) (Hautzinger et al., 1995), Action Regulating Emotion Systems questionnaire (ARES) (Hartig and Moosbrugger, 2003) and Eysenck Personality Inventory (EPI, Form A) (Eggert, 1983). The EPI contains neuroticism and extraversion subscales as well as a "lie" scale (EPI-L). The EPI-L consists of nine items (range: 0-9) assessing response behavior indicative of social desirability (retest-reliability: 0.71, external validity: 0.64) (Eysenck and Eysenck, 1964). Since the results of the current study substantially relied on trait anxiety, which was assessed as self-report based on subjective appraisal by the participant, biases in terms of social desirability cannot be ruled out (Laux et al., 1981). It has been shown that a group of subjects declaring low anxiety is heterogeneous in terms of social desirability proneness, which biases the results when comparing with high anxious subjects (Boor and Schill, 1967). We aimed at identifying those participants who may be prone to social desirable response behavior, indicated by a high score on the EPI-L. Thus, all participants with values of 7 and higher on the EPI-L (corresponding to the upper third of the theoretical range of the scale) were not considered for statistical analysis. Applying this criterion, two subjects (with values of 7 and 8, respectively) were excluded. Of note, these participants had STAI values of 29 and 22, respectively, thus declaring low trait anxiety on the STAI (theoretical range of the STAI: 20-80). Mean (standard deviation) "lie" score of the remaining subjects was 2.8 (1.5). The final sample consisted of 32 subjects (18 female, 14 male). One person was left-handed, all other were right-handed according to self-report and the Annett questionnaire (Annett, 1970). Written informed consent was obtained from all participants. Subjects were compensated for their participation. The study was approved by the cantonal ethics committee (Zurich) and conforms to the Helsinki Declaration.

Magnetic resonance imaging data acquisition

For each participant, one diffusion- and one T1-weighted scan were obtained. Scans were acquired on a 3-T Philips Ingenia whole-body scanner (Philips Medical Systems, Best, The Netherlands) equipped with a transmit-receive body coil and a commercial 15-element sensitivity encoding (SENSE) head coil array.

One diffusion-weighted spin-echo echo-planar imaging (EPI) sequence was applied with a spatial resolution of 2.0 x 2.0 x 2.0 mm^3 (matrix: 112 x 112 pixels, 75 slices in transversal plane). Further imaging parameters were: field of view, 224 x 224 mm^2; echo time, 63.1 ms; repetition time, 18,941.2 ms; flip-angle, 90°; SENSE factor, 2. Diffusion was measured along 64 non-collinear directions ($b = 1000$ s/mm^2) preceded by a non-diffusion-weighted volume (reference volume, $b = 0$ s/mm^2). Scan time was about 23 minutes.

One volumetric 3D T1-weighted gradient echo sequence (fast field echo) with a spatial resolution of 0.94 x 0.94 x 1.00 mm^3 (matrix: 256 x 256 pixels, 160 slices in sagittal plane) was applied. Further imaging parameters were: field of view, 240 x 240 mm^2; echo time, 3.7 ms; repetition time, 8.08 ms; flip-angle, 90°; SENSE factor, 1.5. Scan time was about 8 minutes.

Diffusion tensor imaging preprocessing and tractography

Preprocessing was done with FMRIB Software Library (FSL) Version 4.1.8 (Smith et al., 2004) (www.fmrib.ox.ac.uk/fsl), further processing and deterministic tractography was done using Diffusion Toolkit 0.6.1 and TrackVis 0.5.1 (Wang et al., 2007) (www.trackvis.org). Preprocessing and manual tractography were performed exactly as described previously (Baur et al., in press) (shown also in **supplementary methods S1** [9]). Tractography was performed

---

[9] see appendix, supplementary material of Study 3

for the UF and the inferior fronto-occipital fasciculus. The inferior fronto-occipital fasciculus was considered suitable as control tract because it shares a common trajectory with the UF in the frontal lobe (Catani and Thiebaut de Schotten, 2008). As being part of the well-characterized, large association fiber bundles, manual tractography of the UF and inferior fronto-occipital fasciculus is feasible in a standardized manner (Mori et al., 2002). High inter-rater reliability has been demonstrated for both tracts (Wakana et al., 2007; Baur et al., in press).

Segmentation of subcortical structures

Volumetric segmentation was performed using the FreeSurfer image analysis suite (version 5.1.0; surfer.nmr.mgh.harvard.edu). A detailed description is provided in **supplementary methods S2** [10]. Volumes of subcortical structures (amygdala and hippocampus as structures of interest; caudate nucleus as control structure) as well as total intra-cranial volume were extracted for each participant.

Statistical analysis

To control for global volume (total intra-cranial volume), local WM/GM volume was divided by global, intra-cranial volume. As such, relative volume values were obtained for each measure of interest and used for statistical analysis. For our main approach of *dimensional* associations between trait anxiety (as assessed by the STAI) and WM/GM volume, partial correlations were computed including age, sex, and current depression (as assessed by the BDI) as covariates of no interest, using IBM SPSS Statistics (version 19, SPSS Inc, an IBM company, Armonk, NY). The significance level was $\alpha = 0.05$. All *p*-values are reported uncorrected. We applied strictly hypothesis-driven tests for associations between trait anxiety, UF volume, and amygdala/hippocampus volume. The remaining tests referred to WM/GM control structures and to psychometric variables other than trait anxiety to assess discriminant validity. These tests relied on the assumption *not* to detect a significant correlation. Thus, we refrained from correcting for multiple comparisons to avoid Type II error (Lieberman and Cunningham, 2009). For the association between trait anxiety and left UF volume, *p*-value was interpreted one-tailed because of a directional a-priori hypothesis due to previous evidence (Baur et al., in press). All other *p*-values were two-tailed (given non-directional hypotheses).

---

[10] see appendix, supplementary material of Study 3

Results

Demographic, psychometric and global anatomical measures

Demographic and psychometric measures are summarized in **table 1**. STAI values ranged from 25 to 55 (see **supplementary figure S1** [11]). Mean (standard deviation) intra-cranial volume was 1076.6 (118.8) ml. Intra-cranial volume showed significant positive associations (assessed by Pearson bivariate correlation) with volumes of left UF ($r = 0.52$, $p < 0.01$), right UF ($r = 0.61$, $p < 0.001$), left amygdala ($r = 0.39$, $p < 0.05$), and left hippocampus ($r = 0.46$, $p < 0.01$); non-significant positive associations were observed for right amygdala ($r = 0.20$, $p = 0.27$) and right hippocampus ($r = 0.26$, $p = 0.15$). There was no correlation of intra-cranial volume with trait anxiety ($r = 0.08$, $p = 0.68$). For all further statistical analyses, relative values were used for each measure (that is, local WM/GM volume divided by global, intra-cranial volume).

Table 1: Demographic and psychometric measures ($n = 32$)

	mean (SD)	range	association with trait anxiety [a]
age (years)	24.9 (4.6)	20-37	−0.26 (0.159)
education (years)	16.1 (2.9)	12-22	0.07 (0.688)
trait anxiety [b]	39.3 (8.8)	25-55	
depression [c]	5.0 (4.3)	0-16	0.59 (< 0.001)
anxiety sensitivity [d]	19.0 (8.1)	5-37	0.36 (0.042)
behavioral inhibition [e]	23.8 (5.3)	14-36	0.66 (< 0.001)
neuroticism [f]	9.5 (4.6)	2-20	0.74 (< 0.001)
extraversion [g]	11.3 (3.7)	5-19	−0.23 (0.209)

[a] using Pearson bivariate correlations, shown are r-values (p-values in brackets); [b] according to the Spielberger State-Trait Anxiety Inventory (STAI), trait section; [c] according to the Beck Depression Inventory (BDI); [d] according to the Anxiety Sensitivity Index 3 (ASI-3); [e] according to the BIS subscale of the Action Regulating Emotional Systems (ARES) questionnaire; [f] according to the Eysenck Personality Inventory (EPI), neuroticism subscale; [g] according to the Eysenck Personality Inventory (EPI), extraversion subscale

Associations of trait anxiety with white matter and grey matter volumes

Correlations are summarized in **figure 2**.

Left hemisphere: Trait anxiety was negatively correlated with UF volume ($r = −0.35$, $p = 0.03$), but not with volume of the inferior fronto-occipital fasciculus ($r = 0.01$, $p = 0.94$). Regarding subcortical GM, trait anxiety was positively correlated with amygdala volume ($r = $

---

[11] see appendix, supplementary material of Study 3

0.37, $p = 0.048$). The correlation for the hippocampus was marginally significant, conversely, there was no significant correlation for the caudate nucleus (see **table 2**). UF volume correlated negatively with amygdala volume ($r = -0.40$, $p = 0.03$), but not significantly with the hippocampus or caudate nucleus (see **table 3**). Scatter plots are shown in **supplementary figure S2** [12]. Importantly, results did not change qualitatively when excluding the left-handed subject and the three subjects reporting a history of psychotherapeutic treatment. Correlations of fractional anisotropy, a measure indicative of fiber organization and integrity, for the left UF with trait anxiety and amygdala volume as well as UF volume are presented in **supplementary table S1** [12]. These correlations were similar to those observed for volume.

Right hemisphere: Except for the hippocampus, no significant correlations of WM or GM volumes with trait anxiety (see **table 2**) were observed. Moreover, there was no UF-amygdala volumetric association in the right hemisphere (see **table 3**).

To test discriminant validity of associations with trait anxiety, we additionally assessed correlations with depression, anxiety sensitivity, behavioral inhibition, and neuroticism for each WM/GM volume of interest on an explorative level. Here, no associations were found in both hemispheres (shown in **supplementary table S2** [12]). Post-hoc exploratory tests did not show significant correlations with trait anxiety for the remaining subcortical volumes (**supplementary table S3** [12]). Absolute and relative volumes of all examined WM and GM structures are listed in **supplementary table S4** [12].

---

[12] see appendix, supplementary material of Study 3

Table 2: Associations between white matter/grey matter volumes and trait anxiety

hemisphere	brain tissue	structure	association with trait anxiety [a]
left	*white matter*	uncinate fasciculus	**−0.35 (0.030)** [b]
		inferior fronto-occipital fasciculus	0.01 (0.939)
	grey matter	amygdala	**0.37 (0.048)**
		hippocampus	*0.34 (0.075)*
		caudate	0.09 (0.639)
right	*white matter*	uncinate fasciculus	−0.15 (0.449)
		inferior fronto-occipital fasciculus	−0.04 (0.821)
	grey matter	amygdala	0.28 (0.144)
		hippocampus	**0.44 (0.018)**
		caudate	0.07 (0.710)

[a] using partial correlations with age, sex, and depression as covariates of no interest, shown are *r*-values (*p*-values in brackets, significant results indicated in bold, marginally significant results indicated in italics). Note: For uncinate fasciculus, amygdala, and hippocampus, results underlay hypothesis-driven tests; the remaining tests underlay the null hypothesis; [b] one-tailed

Table 3: Associations between white matter and subcortical grey matter volumes [a]

			white matter			
			left		right	
			UF	IFOF	UF	IFOF
grey matter	left	amygdala	**−0.40**	0.24		
		hippocampus	0.01	0.20		
		caudate	−0.16	0.11		
	right	amygdala			0.07	0.25
		hippocampus			0.12	0.17
		caudate			*−0.32*	−0.05

[a] using partial correlations with age, sex, and depression as covariates of no interest, shown are *r*-values (significant correlations at $p < 0.05$ indicated in bold, marginally significant correlations at $p < 0.10$ indicated in italics)

UF: uncinate fasciculus, IFOF: inferior fronto-occipital fasciculus

Note: Significant/marginally significant results correspond to uncorrected *p*-values. Except of for UF-amygdala volumetric associations, tests underlay the null hypothesis.

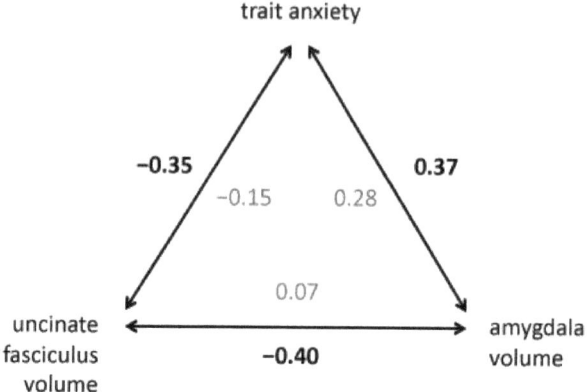

Figure 2: **Triangular association between trait anxiety, uncinate fasciculus volume, and amygdala volume**
Bold (outer) values indicate significant partial correlations (controlling for age, sex, and depression) for the left hemisphere; grey (inner) values indicate respective non-significant correlations for the right hemisphere.

Discussion

This study's focus was on the associations between anxiety and the volumes of the UF, amygdala, and hippocampus as well as on the associations between UF and amygdala. In our non-clinical sample, trait anxiety correlated negatively with UF volume and positively with amygdala volume. In addition, the volumes of UF and amygdala were inversely correlated with each other. These effects were prominent in the left hemisphere and independent of global brain volume, age, sex, and current depression.

In a previous study of our group, we demonstrated reduced left UF volume in patients suffering from social anxiety disorder compared to healthy control subjects using the same deterministic tractography approach (Baur et al., in press). The main goal of the present study was to assess the relation between UF volume and trait anxiety in a non-clinical sample. In this particular sample, we identified a negative correlation of the left UF volume with trait anxiety. In our previous study where we examined a clinical sample, we demonstrated reduced left UF volume in pathologically anxious patients (Baur et al., in press). Both studies demonstrate that anxiety is negatively related to left UF volume, with high anxiety associated with smaller UF volumes. Thus, both studies using entirely different samples point into the same direction and emphasize the pivotal role of the left UF volume in the modulation of anxiety. What may the negative correlation between UF volume and trait anxiety point to? The left UF volume might indicate the structural prerequisites for the efficiency of left amygdala-orbitofrontal functional interactions. These interactions underlie reappraisal as an emotion regulation process (e.g. Kanske et al., 2011) and are critical for reappraisal-initiated reduction of negative affects (Banks et al., 2007). Moreover, changed cross-correlated hemodynamic responses in amygdala-orbitofrontal areas have been demonstrated in pathological anxiety (Liao et al., 2010; Hahn et al., 2011). Thus, our result supports the notion that specific features of structural and functional connectivity within the amygdala-orbitofrontal network are crucial in modulating and possibly determining individual anxiety (Kim et al., 2011). The UF may be considered the main facilitator of signal propagation from and to the amygdala in interactions with the orbitofrontal cortex. It is a "limbic" tract (Catani and Thiebaut de Schotten, 2008; Pugliese et al., 2009) and, in fact, has been described in postmortem and non-human primate studies to innervate the amygdala (Klingler and Gloor, 1960; Ebeling and von Cramon, 1992; Petrides and Pandya, 2007). Yet, further areas also interlinked by the UF may be of interest as well with respect to the present results. The anterior temporal cortex/temporal pole has been related to anxiety (Reiman et al., 1989) and might serve as a relay station between frontal cortex and amygdala, given also strong (possibly UF-independent) projections from the temporal pole to the amygdala (McDonald, 1998; Bach et al., 2011). The present approach used deterministic tractography of well-characterized fiber bundles according to standardized protocols, encompassing the UF *as a whole*. Thus, specific subparts of the UF could not be delineated with this approach. Further studies using more sophisticated approaches could try to disentangle UF fiber subbundles connecting the amygdala and the orbitofrontal cortex (Bach et al., 2011; Saygin et al., 2011). As a supplemental and commonly assessed measure, fractional anisotropy of the left UF was similarly correlated with trait anxiety as compared to volume (**supplementary table S1** [13]). In addition, UF volume and fractional anisotropy were directly positively correlated. It was, however, not within the scope of the present study to explicitly contrast these two measures of WM connectivity. Taken together, the present data rather suggest that UF volume is a measure of interest beside fractional anisotropy.

A further goal of the present study was to examine whether the left UF volume correlates with volume measures of adjacently located brain areas in subcortical GM. In this context we identified, to the best of our knowledge for the first time, that the left-sided UF volume negatively correlated with the left-sided amygdala volume. Beside the relationship between

---

[13] see appendix, supplementary material of Study 3

left UF volume and trait anxiety, we also identified a significant relationship between left amygdala volume and trait anxiety. In addition to our study, there are three studies that have examined the relationship between amygdala volume and trait anxiety so far, with different findings. Two of them reported negative correlations (Spampinato et al., 2009; Blackmon et al., 2011), one no correlation (Kuhn et al., 2011), and our study reports a positive correlation. The reason for these discrepant findings is currently difficult to disentangle due to sample (age etc.) and methodological differences (voxel-based morphometry vs. automatic subcortical segmentation). Further studies are needed to examine the influences that are modulating the amygdala-anxiety relationship. From an intuitive point of view, a positive correlation between amygdala volume and trait anxiety may be plausible: Arguing that a larger amygdala in strong trait-anxious subjects may reflect increased usage would be in line with a couple of studies showing a positive association between amygdala activation and trait anxiety (Etkin et al., 2004; Stein et al., 2007; Carlson et al., 2011; Sehlmeyer et al., 2011). Beside the amygdala, the right hippocampus was significantly and the left hippocampus was trend-wise positively correlated with trait anxiety in the present study (see also Rusch et al., 2001). The anterior hippocampus has been implicated in anxiety-related behavior (Bannerman et al., 2004), as it is also closely connected to the amygdala. As such, our results support the view that the amygdala-hippocampus *complex* is linked to trait anxiety.

The reasons for these specific volumetric associations of the UF, amygdala and hippocampus with trait anxiety as well as between UF and amygdala are unclear so far. However, two possible interpretations can be offered here: First, it could be that the specific structural features we uncovered here are the anatomical prerequisites for inefficient and hyper-responsive anxiety behavior in these subjects. In other words, if this explanation is true, the extent of anxiety-related reactions relies on the specific anatomical features of the UF-amygdala complex. The second interpretation relates more to experience-related structural modifications of the UF-amygdala complex. It could be that in high anxious subjects, the amygdala is more often activated, at the end causing a use-dependent increase of the amygdala volume. Due to this increase, the preponderance to automatically react with anxious behavior increases with less influence from the frontal cortex, causing a decrease of UF volume. Which of these explanations hold true cannot be determined with this experimental design. Longitudinal studies are needed to decide whether a kind of genetically/biologically defined anatomical prerequisite results in anxious reactions or whether frequent fear and anxious feelings modify the structure of involved brain areas.

The results of the present study may also raise the question of laterality with regard to volumetric associations between UF, amygdala, and trait anxiety. The fact that these associations were prominent in the left hemisphere may prompt further investigations to directly address the issue of a possible lateralized involvement of these volumes implicated in limbic emotional circuits associated with anxiety.

Limitations

A major finding of this study is based on self-report data by the participants and it cannot be excluded that there are biases in terms of social desirability. However, the STAI is widely used to measure trait anxiety and there is currently no objective method available to quantify trait anxiety. Due to the reliability of around 0.8 of the STAI (Laux et al., 1981) and the assumed reliability of our anatomical measurements of approximately 0.9 (Baur et al., in press), a maximum structure-behavior relationship of $r_{\text{structure-anxiety, measured}} = r_{\text{structure-anxiety, real}} * \sqrt{0.8*0.9}$ (Vul et al., 2009) can principally be obtained with the applied methods. Thus we are pretty sure that the captured correlations are in the range one could anticipate.

A further limitation is the cross-sectional nature of this study. Thus, we are not in the position to explain our findings being due to experience-driven or biologically-/genetically-driven influences. Longitudinal studies are thus needed to examine the development of anxiety-structure relationships more precisely.

Conclusions

The left-sided UF-amygdala complex is strongly related to anxiety, even in non-pathological, non-clinical subjects. This study is in line with a previous study that has identified exactly the same quality of relationship between UF anatomical features and anxiety in a pathologically anxious sample. Taken together, these and other studies support the notion that the UF-amygdala complex is pivotal for the control of trait anxiety.

Acknowledgments and funding

We thank Roger Luechinger for technical assistance with magnetic resonance imaging and Lui Unterrassner for his contribution to data analysis. The study was funded by a grant of the University of Zurich to LJ.

2.5 Author contributions

Study 1

Volker Baur, Jürgen Hänggi, Uwe Herwig, and Annette B. Brühl designed the study. Michael Rufer, Aba Delsignore, Uwe Herwig, and Annette B. Brühl were responsible for patient recruitment, diagnosis and clinical assessment/testing. Volker Baur, Jürgen Hänggi, Lutz Jäncke, and Annette B. Brühl undertook the statistical analysis. All authors contributed to the discussion and interpretation of the results. Volker Baur wrote the first draft of the manuscript. All authors contributed to and have approved the final manuscript.

Study 2

Volker Baur, Jürgen Hänggi, and Lutz Jäncke designed the study. Michael Rufer, Aba Delsignore, Uwe Herwig, and Annette B. Brühl were responsible for patient recruitment, diagnosis and clinical assessment/testing. Volker Baur, Jürgen Hänggi, Tanja Eberle, and Lutz Jäncke undertook the statistical analysis. All authors contributed to the discussion and interpretation of the results. Volker Baur wrote the first draft of the manuscript. All authors contributed to and have approved the final manuscript.

Study 3

All authors designed the study, undertook the statistical analysis, and contributed to the discussion/interpretation of the results. Volker Baur wrote the first draft of the manuscript. All authors contributed to and have approved the final manuscript.

3 General discussion

3.1 Synopsis

So far, WM connectivity has poorly been characterized with regard to the pathophysiology of SAD. Phan and colleagues were the first to show WM alterations localized in the UF in patients suffering from SAD (Phan et al., 2009). The present studies underline a role of the UF in the pathophysiology of SAD, both from a micro-/mesoscopic (Study 1) and a more large-scale view (Study 2). Reduced FA and volume of the left UF may suggest a connectional deficit between the areas of the frontal and temporal lobe interconnected by the UF. The UF is a main pathway between the orbitofrontal cortex and amygdala. A large body of neuroimaging studies has established a main role of the amygdala in the pathophysiology of SAD (Cannistraro and Rauch, 2003; Etkin and Wager, 2007). Amygdalar hyperactivitiy in SAD can be seen as an exaggerated attentional and emotional response to potential threat (e.g., Stein et al., 2002). The orbitofrontal cortex may be implicated in SAD pathophysiology as well (Goldin et al., 2009). In addition, alterations of orbitofrontal-amygdala functional connectivity in SAD have been demonstrated (Liao et al., 2010; Hahn et al., 2011). Orbitofrontal-amygdala interactions may underlie top-down inhibition of exaggerated emotional responding (Ochsner and Gross, 2005) and the integration of emotion, cognition and motivation (Dolan, 2007) – processes that are likely to be disturbed in anxiety disorders in terms of deficient emotion regulation (Salters-Pedneault et al., 2006) and unbalanced approach-avoidance behavior (Aupperle and Paulus, 2010). Since the UF reflects an anatomical basis for orbitofrontal-amygdala interactions, alterations of the UF are smoothly in line with pathophysiologic models of SAD.

Study 3 was conducted in healthy subjects and qualitatively confirms findings from Study 2 with regard to the direction of volumetric UF-anxiety associations (negative relationship). It can thus be considered – on the one hand – a validation study for newly applied UF volumetric measures in SAD. On the other hand, it may be seen as an extension study, generalizing possible implications of UF anatomical features in terms of anxiety: The UF might be crucial for anxiety-related mechanisms in general, beyond the scope of SAD-specific, pathological processes. This is in line with Kim and Whalen who demonstrated in healthy subjects an association between FA in a small part of the UF and trait anxiety (Kim and Whalen, 2009). Taken together, UF alterations might underlie a general proneness to exaggerated anxiety. Notably and in further support of this model, amygdalar and orbitofrontal morphologic intercorrelations and respective relations to trait anxiety have previously been demonstrated in healthy subjects (Blackmon et al., 2011). Here, the amygdala was volumetrically intercorrelated with the UF in Study 3. Hereby, and in line with previous studies, theoretical models, and known anatomy, attention is drawn to the UF-amygdala *complex*. It remains to be determined in which way UF morphology is linked to orbitofrontal GM morphologic measures such as cortical volume or thickness. This might establish a purely anatomy-based concept of an orbitofrontal-UF-amygdala entity in the framework of anxiety. Some further research questions that arise from or cannot be answered by the present three studies are discussed below and may stimulate future investigations.

3.2 Open questions

1) Are associations between UF measures and anxiety lateralized?

Overall, the present studies found in two independent samples statistically prominent effects in the left – not right – UF regarding associations with SAD or trait anxiety. The left-sided effect is in line with Kim and Whalen (Kim and Whalen, 2009), but in contrast to Phan and colleagues (Phan et al., 2009). In generalized anxiety disorder, UF FA alterations have been identified in the left hemisphere (Hettema et al., 2012). Taking into account studies in major depression, which is often associated with high trait anxiety and comorbid with anxiety disorders (Chambers et al., 2004), results become more inconsistent: Both left-sided (Taylor et al., 2007) and right-sided (Cullen et al., 2010; Hettema et al., 2012) effects in the UF have been demonstrated. Analyses that explicitly focus on hemispheric asymmetry with regard to associations between UF and anxiety-related processes are needed. Furthermore, examinations that include patients with depression may distinguish a subtype of the disorder with prominent anxiety symptoms.

2) Is the UF a WM target structure for psychiatric neuroscience in general?

Beside anxiety disorders and depression, alterations of UF FA have also been shown in schizophrenia (e.g., Kubicki et al., 2002) and bipolar disorder (e.g., Versace et al., 2008). This suggests a more basic role of the UF for the pathophysiology of psychiatric disorders. The following notions support the hypothesis that the UF reflects a basic structural correlate for the vulnerability to or the severity of a psychiatric disorder:

(i) The UF interconnects critical areas involved in the emergence, interpretation and regulation of emotions (amygdala, orbitofrontal cortex) (Ghashghaei et al., 2007; Petrides and Pandya, 2007). Altered emotional processing is present across different psychiatric disorders (Kring and Bachorowski, 1999).

(ii) During childhood and adolescence, FA of the UF linearly increases with age (Eluvathingal et al., 2007; Hasan et al., 2009), rendering this time period to a particularly sensible phase of UF plasticity. Many psychiatric disorders are associated with negative experiences in early life (Heim and Nemeroff, 2002; Weber et al., 2008). One study demonstrated reduced FA of the left UF in children with early-life socio-emotional deprivation (Eluvathingal et al., 2006). Interactions between maturation of the UF, presence of stressors, and the functions of the critical areas the UF interconnects might reflect a susceptibility mechanism for the development of a psychiatric disorder.

Thus, the UF may be of particular interest for affective and psychiatric neuroscience. Comparative studies across different psychiatric disorders are needed to characterize the role of the UF for psychiatric disorders in general. Furthermore, the plastic changes of the UF in the course of a psychiatric disorder should be examined in longitudinal studies, also with regard to psychotherapeutic interventions.

3) May alterations of the UF represent an endophenotype for affective disorders?

The serotonin transporter (5-HTT) is pivotal for serotonin reuptake to the presynaptic neuron. Individuals with a genetic predisposition for reduced 5-HTT number and/or efficiency (5-HTTLPR polymorphism) are at increased risk for an affective disorder (Caspi et al., 2003). Importantly, 5-HTTLPR polymorphism has been related to reduced FA of the frontal part of the left UF (Pacheco et al., 2009). The studies of the present thesis provide a link between measures of the left UF and anxiety. Together with the finding by Pacheco and colleagues, this is well in line with the previously established associations of 5-HTTLPR polymorphism with increased anxiety, increased amygdala activity, and decreased amygdala-prefrontal functional connectivity (Hariri and Holmes, 2006). Alterations of the UF might thus represent an endophenotype for mood and anxiety disorders.

4) Which WM structures may be of interest in terms of anxiety beyond the UF?

Among the large association fiber bundles, the UF seems to be a main target for examinations of anxiety-related associations. The inferior fronto-occipital fasciculus has similar innervation sites in the frontal lobe, but did not show as prominent relations to SAD and trait anxiety as did the UF in Studies 2 and 3. Beyond the large association fiber bundles, future investigations should also focus on WM between specific and circumscribed GM areas. The insula is a cortical integrative area implicated in anxiety and the pathophysiology of anxiety disorders (Etkin and Wager, 2007). Delineating an insula-amygdala pathway and assessing its relations to indicators of anxiety would be a complementary investigation.

5) Volume vs. FA: two sides of one coin or differential predictive value?
In the present studies, associations of both FA and volume of the UF with SAD and trait anxiety were examined. In this context, tract volume was a newly applied measure. Studies 2 and 3 strongly suggest a significance of volumetric measures beside the commonly applied FA measures. Direct (statistical) comparison between volume- and FA-based effects was, however, not applied and should be subject to further studies.[14] Unless more is known about the direct relationship between volume and FA, both measures can be assessed simultaneously (see also Phillips et al., 2011). They may complement one another, as respective implications of FA and volumetric measures may also refer to different spatial scales. Whereas FA is determined voxel-wise – thus operating on a spatial resolution of about 8 mm^3, volume can only be obtained for the tract as a whole when using standard protocols for tractography. FA is generally seen as a "micro-" or "mesoscopic" measure and has the potential to uncover specific sites of altered WM coherence within a single tract. Conversely, the volumetric measure cannot distinguish individual parts of the bundle. Its interpretation may rather be on a large-scale, system level, mainly referring to the amount of WM between given GM areas. In case of the UF, increased volume would mean increased WM presence between amygdala and orbitofrontal cortex, which in turn may reflect an increased capacity for cognitive-emotional interactions.

6) Which are the functional correlates of UF measures?
Complementary examinations using functional MRI should relate tract-specific measures of the UF to the activity of innervated GM areas such as the orbitofrontal cortex, temporal pole and amygdala. This would further validate tract-specific measures like volume, and contribute to a broader understanding of the role of the UF. Activity of UF-innervated GM areas could especially be examined in the context of emotion regulation. Functional connectivity between orbitofrontal cortex and amygdala is crucial for cognitive reappraisal of negative stimuli (Banks et al., 2007; Kanske et al., 2011). UF volume might be linked to these particular activation profiles during an emotion regulation task. It is likely that the UF reflects the main pathway mediating successful emotion regulation, e.g. by top-down inhibition. A recent study

---

[14] see also the discussion section of Study 2

provides first evidence that UF anatomical features are related to the activity of innervated areas: FA in the left UF correlated positively with functional connectivity between amygdala and frontal cortex in the resting state (Steffens et al., 2011). In the framework of the specific functions the UF may facilitate, future work also has to integrate evidence of an involvement of the UF in language and memory (McDonald et al., 2008; Agosta et al., 2010; Papagno et al., 2011).

3.3 Limitations of the present studies

Beside the study-specific limitations discussed in the respective papers, some general restrictive aspects of DTI are also to be mentioned. FA reflects fiber directedness and is interpreted as measure of WM coherence and integrity (Beaulieu, 2009). It is the most widely used DTI measure and is seen as indicator of WM architecture in pathological or different developmental states (Mori and Zhang, 2006) and of WM plasticity (e.g., Jancke et al., 2009; Hanggi et al., 2010). However, it is not straightforward to interpret altered FA values in an experimental group, for instance, patients with SAD as examined here. It is important to keep in mind that multiple microstructural scenarios – such as demyelination, fiber loss, and crossing fibers – can lead to an observed reduction in FA (Mori and Zhang, 2006). Deterministic tractography (applied in Study 2 and 3) relies on the principal diffusion direction represented by V_1.[15] This may lead to a conflict if fibers are crossing within a voxel, since one single fiber direction for a voxel has to be determined for the tractography process. Probabilistic tractography (Behrens et al., 2003) and diffusion spectrum imaging (Wedeen et al., 2008) can overcome this limitation. In contrast to the noted limitations, the capacity of the non-invasive, in-vivo examination of WM structure is the striking strength of DTI (Behrens and Jbabdi, 2009).

3.4 Concluding remarks

Previous studies of neuroanatomical and morphologic bases of anxiety and anxiety disorders mainly focused on GM. The present studies are complementary by adding a WM structure of significance, the UF, for anxiety-related processes. They support the importance of the UF for the pathophysiology of SAD and help establish its role in general anxiety. Specific anatomical features of the UF might be linked to the anxiety continuum ranging from normal to

---

[15] see methodological introduction

pathological. This bears important clinical implications, but may also push forward our understanding of the way anxiety evolves from cerebral processes.

4 References

Ackenheil M, Stotz-Ingenlath G, Dietz-Bauer R, Vossen A (1999) M.I.N.I. Mini International Neuropsychiatric Interview, German Version 5.0.0 DSM-IV. München: Psychiatrische Universitätsklinik München.

Adolphs R (2003a) Is the human amygdala specialized for processing social information? Ann N Y Acad Sci 985:326-340.

Adolphs R (2003b) Cognitive neuroscience of human social behaviour. Nat Rev Neurosci 4:165-178.

Agosta F, Henry RG, Migliaccio R, Neuhaus J, Miller BL, Dronkers NF, Brambati SM, Filippi M, Ogar JM, Wilson SM, Gorno-Tempini ML (2010) Language networks in semantic dementia. Brain 133:286-299.

Akirav I, Maroun M (2007) The role of the medial prefrontal cortex-amygdala circuit in stress effects on the extinction of fear. Neural Plast 2007:30873.

Alexander GE, DeLong MR, Strick PL (1986) Parallel organization of functionally segregated circuits linking basal ganglia and cortex. Annu Rev Neurosci 9:357-381.

American Psychiatric Association (1994) Diagnostic and statistical manual of mental disorders, 4th edition Edition. Washington, DC: American Psychiatric Press.

Annett M (1970) A classification of hand preference by association analysis. Br J Psychol 61:303-321.

Aupperle RL, Paulus MP (2010) Neural systems underlying approach and avoidance in anxiety disorders. Dialogues Clin Neurosci 12:517-531.

Bach DR, Behrens TE, Garrido L, Weiskopf N, Dolan RJ (2011) Deep and superficial amygdala nuclei projections revealed in vivo by probabilistic tractography. J Neurosci 31:618-623.

Baker SL, Heinrichs N, Kim HJ, Hofmann SG (2002) The liebowitz social anxiety scale as a self-report instrument: a preliminary psychometric analysis. Behav Res Ther 40:701-715.

Banks SJ, Eddy KT, Angstadt M, Nathan PJ, Phan KL (2007) Amygdala-frontal connectivity during emotion regulation. Soc Cogn Affect Neurosci 2:303-312.

Bannerman DM, Rawlins JN, McHugh SB, Deacon RM, Yee BK, Bast T, Zhang WN, Pothuizen HH, Feldon J (2004) Regional dissociations within the hippocampus--memory and anxiety. Neurosci Biobehav Rev 28:273-283.

4 References

Basser PJ, Pierpaoli C (1996) Microstructural and physiological features of tissues elucidated by quantitative-diffusion-tensor MRI. J Magn Reson B 111:209-219.

Basser PJ, Özarslan E (2009) Introduction to diffusion MR. In: Diffusion MRI: From quantitative measurement to in vivo neuroanatomy (Johansen-Berg H, Behrens TEJ, eds). Amsterdam: Academic Press.

Baur V, Hanggi J, Rufer M, Delsignore A, Jancke L, Herwig U, Bruhl AB (2011) White matter alterations in social anxiety disorder. J Psychiatr Res 45:1366-1372.

Baur V, Bruhl AB, Herwig U, Eberle T, Rufer M, Delsignore A, Jancke L, Hanggi J (in press) Evidence of frontotemporal structural hypoconnectivity in social anxiety disorder: A quantitative fiber tractography study. Hum Brain Mapp.

Beaulieu C (2009) The biological basis of diffusion anisotropy. In: Diffusion MRI: From quantitative measurement to in vivo neuroanatomy (Johansen-Berg H, Behrens TEJ, eds). Amsterdam: Academic Press.

Beck AT, Ward CH, Mendelson M, Mock J, Erbaugh J (1961) An inventory for measuring depression. Arch Gen Psychiatry 4:561-571.

Behrens TE, Woolrich MW, Jenkinson M, Johansen-Berg H, Nunes RG, Clare S, Matthews PM, Brady JM, Smith SM (2003) Characterization and propagation of uncertainty in diffusion-weighted MR imaging. Magn Reson Med 50:1077-1088.

Behrens TEJ, Jbabdi S (2009) MR Diffusion Tractography. In: Diffusion MRI: From quantitative measurement to in vivo neuroanatomy (Johansen-Berg H, Behrens TEJ, eds). Amsterdam: Academic Press.

Berkowitz RL, Coplan JD, Reddy DP, Gorman JM (2007) The human dimension: how the prefrontal cortex modulates the subcortical fear response. Rev Neurosci 18:191-207.

Bishop SJ (2007) Neurocognitive mechanisms of anxiety: an integrative account. Trends Cogn Sci 11:307-316.

Bishop SJ (2009) Trait anxiety and impoverished prefrontal control of attention. Nat Neurosci 12:92-98.

Blackmon K, Barr WB, Carlson C, Devinsky O, Dubois J, Pogash D, Quinn BT, Kuzniecky R, Halgren E, Thesen T (2011) Structural evidence for involvement of a left amygdala-orbitofrontal network in subclinical anxiety. Psychiatry Res 194:296-303.

Boor M, Schill T (1967) Digit symbol performance of subjects varying in anxiety and defensiveness. J Consult Psychol 31:600-603.

Brett M, Anton J, Valabregue R, Poline J (2002) Region of interest analysis using an SPM toolbox [abstract] Presented at the 8th International Conference on Functional

Mapping of the Human Brain, June 2-6, 2002, Sendai, Japan Available on CD-ROM in Neuroimage, Vol 16, No 2.

Bruhl AB, Rufer M, Delsignore A, Kaffenberger T, Jancke L, Herwig U (2011) Neural correlates of altered general emotion processing in social anxiety disorder. Brain Res 1378:72-83.

Cannistraro PA, Rauch SL (2003) Neural circuitry of anxiety: evidence from structural and functional neuroimaging studies. Psychopharmacol Bull 37:8-25.

Carlson JM, Greenberg T, Rubin D, Mujica-Parodi LR (2011) Feeling anxious: anticipatory amygdalo-insular response predicts the feeling of anxious anticipation. Soc Cogn Affect Neurosci 6:74-81.

Caspi A, Sugden K, Moffitt TE, Taylor A, Craig IW, Harrington H, McClay J, Mill J, Martin J, Braithwaite A, Poulton R (2003) Influence of life stress on depression: moderation by a polymorphism in the 5-HTT gene. Science 301:386-389.

Castren E (2005) Is mood chemistry? Nat Rev Neurosci 6:241-246.

Catani M, Thiebaut de Schotten M (2008) A diffusion tensor imaging tractography atlas for virtual in vivo dissections. Cortex 44:1105-1132.

Catani M, Howard RJ, Pajevic S, Jones DK (2002) Virtual in vivo interactive dissection of white matter fasciculi in the human brain. Neuroimage 17:77-94.

Chambers JA, Power KG, Durham RC (2004) The relationship between trait vulnerability and anxiety and depressive diagnoses at long-term follow-up of Generalized Anxiety Disorder. J Anxiety Disord 18:587-607.

Cullen KR, Klimes-Dougan B, Muetzel R, Mueller BA, Camchong J, Houri A, Kurma S, Lim KO (2010) Altered white matter microstructure in adolescents with major depression: a preliminary study. J Am Acad Child Adolesc Psychiatry 49:173-183 e171.

Davis M, Whalen PJ (2001) The amygdala: vigilance and emotion. Mol Psychiatry 6:13-34.

Dolan RJ (2007) The human amygdala and orbital prefrontal cortex in behavioural regulation. Philos Trans R Soc Lond B Biol Sci 362:787-799.

Ebeling U, von Cramon D (1992) Topography of the uncinate fascicle and adjacent temporal fiber tracts. Acta Neurochir (Wien) 115:143-148.

Edgar JM, Griffiths IR (2009) White matter structure: a microscopist's view. In: Diffusion MRI: From quantitative measurement to in vivo neuroanatomy (Johansen-Berg H, Behrens TEJ, eds). Amsterdam: Academic Press.

Eggert D (1983) Eysenck-Persönlichkeits-Inventar E-P-I. Handanweisung für die Durchführung und Auswertung. Göttingen, Toronto, Zürich: Hogrefe.

4 References

Eickhoff SB, Stephan KE, Mohlberg H, Grefkes C, Fink GR, Amunts K, Zilles K (2005) A new SPM toolbox for combining probabilistic cytoarchitectonic maps and functional imaging data. Neuroimage 25:1325-1335.

Eluvathingal TJ, Hasan KM, Kramer L, Fletcher JM, Ewing-Cobbs L (2007) Quantitative diffusion tensor tractography of association and projection fibers in normally developing children and adolescents. Cereb Cortex 17:2760-2768.

Eluvathingal TJ, Chugani HT, Behen ME, Juhasz C, Muzik O, Maqbool M, Chugani DC, Makki M (2006) Abnormal brain connectivity in children after early severe socioemotional deprivation: a diffusion tensor imaging study. Pediatrics 117:2093-2100.

Etkin A, Wager TD (2007) Functional neuroimaging of anxiety: a meta-analysis of emotional processing in PTSD, social anxiety disorder, and specific phobia. Am J Psychiatry 164:1476-1488.

Etkin A, Prater KE, Schatzberg AF, Menon V, Greicius MD (2009) Disrupted amygdalar subregion functional connectivity and evidence of a compensatory network in generalized anxiety disorder. Arch Gen Psychiatry 66:1361-1372.

Etkin A, Klemenhagen KC, Dudman JT, Rogan MT, Hen R, Kandel ER, Hirsch J (2004) Individual differences in trait anxiety predict the response of the basolateral amygdala to unconsciously processed fearful faces. Neuron 44:1043-1055.

Eysenck HJ, Eysenck SBG (1964) Manual of the Eysenck Personality Inventory. London: University of London Press.

Fischl B, Salat DH, Busa E, Albert M, Dieterich M, Haselgrove C, van der Kouwe A, Killiany R, Kennedy D, Klaveness S, Montillo A, Makris N, Rosen B, Dale AM (2002) Whole brain segmentation: automated labeling of neuroanatomical structures in the human brain. Neuron 33:341-355.

Fischl B, van der Kouwe A, Destrieux C, Halgren E, Segonne F, Salat DH, Busa E, Seidman LJ, Goldstein J, Kennedy D, Caviness V, Makris N, Rosen B, Dale AM (2004) Automatically parcellating the human cerebral cortex. Cereb Cortex 14:11-22.

Freitas-Ferrari MC, Hallak JE, Trzesniak C, Filho AS, Machado-de-Sousa JP, Chagas MH, Nardi AE, Crippa JA (2010) Neuroimaging in social anxiety disorder: a systematic review of the literature. Prog Neuropsychopharmacol Biol Psychiatry 34:565-580.

Ghashghaei HT, Barbas H (2002) Pathways for emotion: interactions of prefrontal and anterior temporal pathways in the amygdala of the rhesus monkey. Neuroscience 115:1261-1279.

Ghashghaei HT, Hilgetag CC, Barbas H (2007) Sequence of information processing for emotions based on the anatomic dialogue between prefrontal cortex and amygdala. Neuroimage 34:905-923.

Goldin PR, Manber T, Hakimi S, Canli T, Gross JJ (2009) Neural bases of social anxiety disorder: emotional reactivity and cognitive regulation during social and physical threat. Arch Gen Psychiatry 66:170-180.

Ha TH, Kang DH, Park JS, Jang JH, Jung WH, Choi JS, Park JY, Jung MH, Choi CH, Lee JM, Ha K, Kwon JS (2009) White matter alterations in male patients with obsessive-compulsive disorder. Neuroreport 20:735-739.

Hahn A, Stein P, Windischberger C, Weissenbacher A, Spindelegger C, Moser E, Kasper S, Lanzenberger R (2011) Reduced resting-state functional connectivity between amygdala and orbitofrontal cortex in social anxiety disorder. Neuroimage 56:881-889.

Hanggi J, Koeneke S, Bezzola L, Jancke L (2010) Structural neuroplasticity in the sensorimotor network of professional female ballet dancers. Hum Brain Mapp 31:1196-1206.

Hariri AR, Holmes A (2006) Genetics of emotional regulation: the role of the serotonin transporter in neural function. Trends Cogn Sci 10:182-191.

Hartig J, Moosbrugger H (2003) Die "ARES-Skalen" zur Erfassung der individuellen BIS- und BAS-Sensitivität. Zeitschrift für Differentielle und Diagnostische Psychologie 24:293-310.

Hasan KM, Iftikhar A, Kamali A, Kramer LA, Ashtari M, Cirino PT, Papanicolaou AC, Fletcher JM, Ewing-Cobbs L (2009) Development and aging of the healthy human brain uncinate fasciculus across the lifespan using diffusion tensor tractography. Brain Res 1276:67-76.

Hautzinger M, Bailer M, Worall H, Keller F (1995) Beck-Depressions-Inventar (BDI) - Testhandbuch. Bern, Göttingen, Toronto, Seattle: Huber.

Hayasaka S, Nichols TE (2004) Combining voxel intensity and cluster extent with permutation test framework. Neuroimage 23:54-63.

Hayasaka S, Phan KL, Liberzon I, Worsley KJ, Nichols TE (2004) Nonstationary cluster-size inference with random field and permutation methods. Neuroimage 22:676-687.

Heim C, Nemeroff CB (2002) Neurobiology of early life stress: clinical studies. Semin Clin Neuropsychiatry 7:147-159.

Heimer L (2003) A new anatomical framework for neuropsychiatric disorders and drug abuse. Am J Psychiatry 160:1726-1739.

Hettema JM, Kettenmann B, Ahluwalia V, McCarthy C, Kates WR, Schmitt JE, Silberg JL, Neale MC, Kendler KS, Fatouros P (2012) Pilot multimodal twin imaging study of generalized anxiety disorder. Depress Anxiety 29:202-209.

Hirsch CR, Clark DM (2004) Information-processing bias in social phobia. Clin Psychol Rev 24:799-825.

Hua K, Zhang J, Wakana S, Jiang H, Li X, Reich DS, Calabresi PA, Pekar JJ, van Zijl PC, Mori S (2008) Tract probability maps in stereotaxic spaces: analyses of white matter anatomy and tract-specific quantification. Neuroimage 39:336-347.

Huang H, Zhang J, van Zijl PC, Mori S (2004) Analysis of noise effects on DTI-based tractography using the brute-force and multi-ROI approach. Magn Reson Med 52:559-565.

Imfeld A, Oechslin MS, Meyer M, Loenneker T, Jancke L (2009) White matter plasticity in the corticospinal tract of musicians: a diffusion tensor imaging study. Neuroimage 46:600-607.

Jancke L, Koeneke S, Hoppe A, Rominger C, Hanggi J (2009) The architecture of the golfer's brain. PLoS One 4:e4785.

Jefferys D (1997) Social phobia. The most common anxiety disorder. Aust Fam Physician 26:1061, 1064-1067.

Kanske P, Heissler J, Schonfelder S, Bongers A, Wessa M (2011) How to regulate emotion? Neural networks for reappraisal and distraction. Cereb Cortex 21:1379-1388.

Kawashima T, Nakamura M, Bouix S, Kubicki M, Salisbury DF, Westin CF, McCarley RW, Shenton ME (2009) Uncinate fasciculus abnormalities in recent onset schizophrenia and affective psychosis: a diffusion tensor imaging study. Schizophr Res 110:119-126.

Kessler RC, McGonagle KA, Zhao S, Nelson CB, Hughes M, Eshleman S, Wittchen HU, Kendler KS (1994) Lifetime and 12-month prevalence of DSM-III-R psychiatric disorders in the United States. Results from the National Comorbidity Survey. Arch Gen Psychiatry 51:8-19.

Kim MJ, Whalen PJ (2009) The structural integrity of an amygdala-prefrontal pathway predicts trait anxiety. J Neurosci 29:11614-11618.

Kim MJ, Loucks RA, Palmer AL, Brown AC, Solomon KM, Marchante AN, Whalen PJ (2011) The structural and functional connectivity of the amygdala: from normal emotion to pathological anxiety. Behav Brain Res 223:403-410.

Klingler J, Gloor P (1960) The connections of the amygdala and of the anterior temporal cortex in the human brain. J Comp Neurol 115:333-369.

4 References

Kring AM, Bachorowski J (1999) Emotions and psychopathology. Cognition and Emotion 13:575-599.

Kubicki M (2010) Neurocognition and white matter imaging: can the relationship be reliably quantified? Am J Psychiatry 167:373-375.

Kubicki M, Westin CF, Maier SE, Frumin M, Nestor PG, Salisbury DF, Kikinis R, Jolesz FA, McCarley RW, Shenton ME (2002) Uncinate fasciculus findings in schizophrenia: a magnetic resonance diffusion tensor imaging study. Am J Psychiatry 159:813-820.

Kuhn S, Schubert F, Gallinat J (2011) Structural correlates of trait anxiety: Reduced thickness in medial orbitofrontal cortex accompanied by volume increase in nucleus accumbens. J Affect Disord 134:315-319.

Laux L, Glanzmann P, Schaffner P, Spielberger CD (1981) Das State-Trait-Angstinventar - Manual. Weinheim: Beltz.

LeDoux JE, Gorman JM (2001) A call to action: overcoming anxiety through active coping. Am J Psychiatry 158:1953-1955.

Leemans A (2006) Modeling and processing of diffusion tensor magnetic resonance images for improved analysis of brain connectivity: University of Antwerp, unpublished dissertation.

Liao W, Qiu C, Gentili C, Walter M, Pan Z, Ding J, Zhang W, Gong Q, Chen H (2010) Altered effective connectivity network of the amygdala in social anxiety disorder: a resting-state FMRI study. PLoS One 5:e15238.

Liao W, Xu Q, Mantini D, Ding J, Machado-de-Sousa JP, Hallak JE, Trzesniak C, Qiu C, Zeng L, Zhang W, Crippa JA, Gong Q, Chen H (2011) Altered gray matter morphometry and resting-state functional and structural connectivity in social anxiety disorder. Brain Res 1388:167-177.

Lieberman MD, Cunningham WA (2009) Type I and Type II error concerns in fMRI research: re-balancing the scale. Soc Cogn Affect Neurosci 4:423-428.

Liebowitz MR (1987) Social phobia. Mod Probl Pharmacopsychiatry 22:141-173.

McDonald AJ (1998) Cortical pathways to the mammalian amygdala. Prog Neurobiol 55:257-332.

McDonald CR, Ahmadi ME, Hagler DJ, Tecoma ES, Iragui VJ, Gharapetian L, Dale AM, Halgren E (2008) Diffusion tensor imaging correlates of memory and language impairments in temporal lobe epilepsy. Neurology 71:1869-1876.

McIntosh AM, Munoz Maniega S, Lymer GK, McKirdy J, Hall J, Sussmann JE, Bastin ME, Clayden JD, Johnstone EC, Lawrie SM (2008) White matter tractography in bipolar disorder and schizophrenia. Biol Psychiatry 64:1088-1092.

Mori S, van Zijl PC (2002) Fiber tracking: principles and strategies - a technical review. NMR Biomed 15:468-480.

Mori S, Zhang J (2006) Principles of diffusion tensor imaging and its applications to basic neuroscience research. Neuron 51:527-539.

Mori S, Crain BJ, Chacko VP, van Zijl PC (1999) Three-dimensional tracking of axonal projections in the brain by magnetic resonance imaging. Ann Neurol 45:265-269.

Mori S, Kaufmann WE, Davatzikos C, Stieltjes B, Amodei L, Fredericksen K, Pearlson GD, Melhem ER, Solaiyappan M, Raymond GV, Moser HW, van Zijl PC (2002) Imaging cortical association tracts in the human brain using diffusion-tensor-based axonal tracking. Magn Reson Med 47:215-223.

Mosing MA, Gordon SD, Medland SE, Statham DJ, Nelson EC, Heath AC, Martin NG, Wray NR (2009) Genetic and environmental influences on the co-morbidity between depression, panic disorder, agoraphobia, and social phobia: a twin study. Depress Anxiety 26:1004-1011.

Ochsner KN, Gross JJ (2005) The cognitive control of emotion. Trends Cogn Sci 9:242-249.

Oechslin MS, Imfeld A, Loenneker T, Meyer M, Jancke L (2009) The plasticity of the superior longitudinal fasciculus as a function of musical expertise: a diffusion tensor imaging study. Front Hum Neurosci 3:76.

Omura K, Todd Constable R, Canli T (2005) Amygdala gray matter concentration is associated with extraversion and neuroticism. Neuroreport 16:1905-1908.

Pacheco J, Beevers CG, Benavides C, McGeary J, Stice E, Schnyer DM (2009) Frontal-limbic white matter pathway associations with the serotonin transporter gene promoter region (5-HTTLPR) polymorphism. J Neurosci 29:6229-6233.

Papagno C, Miracapillo C, Casarotti A, Romero Lauro LJ, Castellano A, Falini A, Casaceli G, Fava E, Bello L (2011) What is the role of the uncinate fasciculus? Surgical removal and proper name retrieval. Brain 134:405-414.

Park HJ, Westin CF, Kubicki M, Maier SE, Niznikiewicz M, Baer A, Frumin M, Kikinis R, Jolesz FA, McCarley RW, Shenton ME (2004) White matter hemisphere asymmetries in healthy subjects and in schizophrenia: a diffusion tensor MRI study. Neuroimage 23:213-223.

Petrides M, Pandya DN (2007) Efferent association pathways from the rostral prefrontal cortex in the macaque monkey. J Neurosci 27:11573-11586.

Phan KL, Orlichenko A, Boyd E, Angstadt M, Coccaro EF, Liberzon I, Arfanakis K (2009) Preliminary evidence of white matter abnormality in the uncinate fasciculus in generalized social anxiety disorder. Biol Psychiatry 66:691-694.

Phillips OR, Clark KA, Woods RP, Subotnik KL, Asarnow RF, Nuechterlein KH, Toga AW, Narr KL (2011) Topographical relationships between arcuate fasciculus connectivity and cortical thickness. Hum Brain Mapp 32:1788-1801.

Pugliese L, Catani M, Ameis S, Dell'Acqua F, Thiebaut de Schotten M, Murphy C, Robertson D, Deeley Q, Daly E, Murphy DG (2009) The anatomy of extended limbic pathways in Asperger syndrome: a preliminary diffusion tensor imaging tractography study. Neuroimage 47:427-434.

Reiman EM, Raichle ME, Robins E, Mintun MA, Fusselman MJ, Fox PT, Price JL, Hackman KA (1989) Neuroanatomical correlates of a lactate-induced anxiety attack. Arch Gen Psychiatry 46:493-500.

Reiss S, Peterson RA, Gursky DM, McNally RJ (1986) Anxiety sensitivity, anxiety frequency and the prediction of fearfulness. Behav Res Ther 24:1-8.

Rodrigo S, Oppenheim C, Chassoux F, Golestani N, Cointepas Y, Poupon C, Semah F, Mangin JF, Le Bihan D, Meder JF (2007) Uncinate fasciculus fiber tracking in mesial temporal lobe epilepsy. Initial findings. Eur Radiol 17:1663-1668.

Rusch BD, Abercrombie HC, Oakes TR, Schaefer SM, Davidson RJ (2001) Hippocampal morphometry in depressed patients and control subjects: relations to anxiety symptoms. Biol Psychiatry 50:960-964.

Salters-Pedneault K, Roemer L, Tull MT, Rucker L, Mennin DS (2006) Evidence of Broad Deficits in Emotion Regulation Associated with Chronic Worry and Generalized Anxiety Disorder. Cogn Ther Res 30:469-480.

Saygin ZM, Osher DE, Augustinack J, Fischl B, Gabrieli JD (2011) Connectivity-based segmentation of human amygdala nuclei using probabilistic tractography. Neuroimage 56:1353-1361.

Schmithorst VJ, Holland SK, Plante E (2011) Diffusion tensor imaging reveals white matter microstructure correlations with auditory processing ability. Ear Hear 32:156-167.

Sehlmeyer C, Dannlowski U, Schoning S, Kugel H, Pyka M, Pfleiderer B, Zwitserlood P, Schiffbauer H, Heindel W, Arolt V, Konrad C (2011) Neural correlates of trait anxiety in fear extinction. Psychol Med 41:789-798.

4 *References*

Sheehan DV, Lecrubier Y, Sheehan KH, Amorim P, Janavs J, Weiller E, Hergueta T, Baker R, Dunbar GC (1998) The Mini-International Neuropsychiatric Interview (M.I.N.I.): the development and validation of a structured diagnostic psychiatric interview for DSM-IV and ICD-10. J Clin Psychiatry 59:22-33.

Smith SM (2002) Fast robust automated brain extraction. Hum Brain Mapp 17:143-155.

Smith SM, Jenkinson M, Woolrich MW, Beckmann CF, Behrens TEJ, Johansen-Berg H, Bannister PR, De Luca M, Drobnjak I, Flitney DE, Niazy RK, Saunders J, Vickers J, Zhang Y, De Stefano N, Brady JM, Matthews PM (2004) Advances in functional and structural MR image analysis and implementation as FSL. NeuroImage 23:S208-S219.

Somers JM, Goldner EM, Waraich P, Hsu L (2006) Prevalence and incidence studies of anxiety disorders: a systematic review of the literature. Can J Psychiatry 51:100-113.

Song SK, Sun SW, Ramsbottom MJ, Chang C, Russell J, Cross AH (2002) Dysmyelination revealed through MRI as increased radial (but unchanged axial) diffusion of water. Neuroimage 17:1429-1436.

Spampinato MV, Wood JN, De Simone V, Grafman J (2009) Neural correlates of anxiety in healthy volunteers: a voxel-based morphometry study. J Neuropsychiatry Clin Neurosci 21:199-205.

Spielberger CD, Gorsuch RL, Lushene RE (1970) State-Trait Anxiety Inventory, Manual for the State-Trait Anxiety Inventory. Palo Alto, CA: Consulting Psychologist Press.

Stangier U, Heidenreich T (2005) Die Liebowitz Soziale Angst Skala (LSAS). Göttingen: Hogrefe.

Steffens DC, Taylor WD, Denny KL, Bergman SR, Wang L (2011) Structural integrity of the uncinate fasciculus and resting state functional connectivity of the ventral prefrontal cortex in late life depression. PLoS One 6:e22697.

Stein MB, Stein DJ (2008) Social anxiety disorder. Lancet 371:1115-1125.

Stein MB, Simmons AN, Feinstein JS, Paulus MP (2007) Increased amygdala and insula activation during emotion processing in anxiety-prone subjects. Am J Psychiatry 164:318-327.

Stein MB, Goldin PR, Sareen J, Zorrilla LT, Brown GG (2002) Increased amygdala activation to angry and contemptuous faces in generalized social phobia. Arch Gen Psychiatry 59:1027-1034.

Sussmann JE, Lymer GK, McKirdy J, Moorhead TW, Munoz Maniega S, Job D, Hall J, Bastin ME, Johnstone EC, Lawrie SM, McIntosh AM (2009) White matter abnormalities in bipolar disorder and schizophrenia detected using diffusion tensor magnetic resonance imaging. Bipolar Disord 11:11-18.

Taylor WD, MacFall JR, Gerig G, Krishnan RR (2007) Structural integrity of the uncinate fasciculus in geriatric depression: Relationship with age of onset. Neuropsychiatr Dis Treat 3:669-674.

Thiebaut de Schotten M, Dell'acqua F, Valabregue R, Catani M (2012) Monkey to human comparative anatomy of the frontal lobe association tracts. Cortex 48:82-96.

Tillfors M, Furmark T, Marteinsdottir I, Fischer H, Pissiota A, Langstrom B, Fredrikson M (2001) Cerebral blood flow in subjects with social phobia during stressful speaking tasks: a PET study. Am J Psychiatry 158:1220-1226.

Turk CL, Heimberg RG, Luterek JA, Mennin DS, Fresco DM (2005) Emotion Dysregulation in Generalized Anxiety Disorder: A Comparison with Social Anxiety Disorder. Cogn Ther Res 29:89-106.

Versace A, Almeida JR, Hassel S, Walsh ND, Novelli M, Klein CR, Kupfer DJ, Phillips ML (2008) Elevated left and reduced right orbitomedial prefrontal fractional anisotropy in adults with bipolar disorder revealed by tract-based spatial statistics. Arch Gen Psychiatry 65:1041-1052.

Vul E, Harris C, Winkielman P, Pashler H (2009) Puzzlingly high correlations in fMRI studies of emotion, personality, and social cognition. Persp Psychol Sci 4:274-290.

Wakana S, Caprihan A, Panzenboeck MM, Fallon JH, Perry M, Gollub RL, Hua K, Zhang J, Jiang H, Dubey P, Blitz A, van Zijl P, Mori S (2007) Reproducibility of quantitative tractography methods applied to cerebral white matter. Neuroimage 36:630-644.

Wang F, Kalmar JH, Edmiston E, Chepenik LG, Bhagwagar Z, Spencer L, Pittman B, Jackowski M, Papademetris X, Constable RT, Blumberg HP (2008a) Abnormal corpus callosum integrity in bipolar disorder: a diffusion tensor imaging study. Biol Psychiatry 64:730-733.

Wang F, Jackowski M, Kalmar JH, Chepenik LG, Tie K, Qiu M, Gong G, Pittman BP, Jones MM, Shah MP, Spencer L, Papademetris X, Constable RT, Blumberg HP (2008b) Abnormal anterior cingulum integrity in bipolar disorder determined through diffusion tensor imaging. Br J Psychiatry 193:126-129.

Wang R, Benner T, Sorensen AG, Wedeen VJ (2007) Diffusion Toolkit: A software package for diffusion imaging data processing and tractography. Proc Intl Soc Mag Reson Med 15:3720.

Weber K, Rockstroh B, Borgelt J, Awiszus B, Popov T, Hoffmann K, Schonauer K, Watzl H, Propster K (2008) Stress load during childhood affects psychopathology in psychiatric patients. BMC Psychiatry 8:63.

Wedeen VJ, Wang RP, Schmahmann JD, Benner T, Tseng WY, Dai G, Pandya DN, Hagmann P, D'Arceuil H, de Crespigny AJ (2008) Diffusion spectrum magnetic resonance imaging (DSI) tractography of crossing fibers. Neuroimage 41:1267-1277.

Worsley KJ, Andermann M, Koulis T, MacDonald D, Evans AC (1999) Detecting changes in nonisotropic images. Hum Brain Mapp 8:98-101.

Wright CI, Williams D, Feczko E, Barrett LF, Dickerson BC, Schwartz CE, Wedig MM (2006) Neuroanatomical correlates of extraversion and neuroticism. Cereb Cortex 16:1809-1819.

Xu J, Potenza MN (2012) White matter integrity and five-factor personality measures in healthy adults. Neuroimage 59:800-807.

Yasmin H, Aoki S, Abe O, Nakata Y, Hayashi N, Masutani Y, Goto M, Ohtomo K (2009) Tract-specific analysis of white matter pathways in healthy subjects: a pilot study using diffusion tensor MRI. Neuroradiology 51:831-840.

Yurgelun-Todd DA, Silveri MM, Gruber SA, Rohan ML, Pimentel PJ (2007) White matter abnormalities observed in bipolar disorder: a diffusion tensor imaging study. Bipolar Disord 9:504-512.

Zhou W, Hou P, Zhou Y, Chen D (2011) Reduced recruitment of orbitofrontal cortex to human social chemosensory cues in social anxiety. Neuroimage 55:1401-1406.

Appendix

Supplementary Material of Study 1

Table S1: Cross-correlations of measures of anxiety and depression in patients with social anxiety disorder

	r	p	n
STAI with LSAS	0.62	< 0.01	25
STAI with BDI	0.57	< 0.01	23
LSAS with BDI	0.37	0.08	23

STAI: Spielberger State-Trait Anxiety Inventory (trait version), LSAS: Liebowitz Social Anxiety Scale, BDI: Beck's Depression Inventory

Table S2: Group comparison of fractional anisotropy (patients with social anxiety disorder < healthy subjects, $p < 0.00001$ voxel-wise uncorrected combined with $p < 0.05$ cluster-extent FWE-corrected) using 3-mm smoothed FA maps

cluster	size (mm^3)	MNI peak coordinates			T (max)
		x	y	z	
UF/IFOF (left)	86	-24	39	-8	6.18
SLF (left)	80	-31	-30	20	5.25
superior cerebellar pedunculus (right)	53	6	-44	-27	6.06
corpus callosum (left)	57	-27	-55	14	5.48

UF: uncinate fasciculus, IFOF: inferior fronto-occipital fasciculus, SLF: superior longitudinal fasciculus

Appendix

Table S3: Medicated vs. medication-free patients with social anxiety disorder: post-hoc analysis of the significant clusters (see table 2). *p*-values at the rightmost column are uncorrected and correspond to ANCOVA for each cluster, contrasting medicated ($n = 9$) vs. medication-free ($n = 16$) patients with social anxiety disorder (including age and trait anxiety as covariates of no interest)

cluster	mean FA		F	p
	medicated	medication-free		
UF/IFOF (left)	0.21 ± 0.03	0.20 ± 0.03	0.34	0.57
SLF (left)	0.38 ± 0.02	0.39 ± 0.03	1.44	0.24

UF: uncinate fasciculus, IFOF: inferior fronto-occipital fasciculus, SLF: superior longitudinal fasciculus; FA: fractional anisotropy

Figure S4: Medicated vs. medication-free patients with social anxiety disorder: post-hoc analysis of the significant clusters (see table 2). Bars represent mean fractional anisotropy (FA) ± standard deviation across both significant clusters of the FA group comparison (patients with social anxiety disorder < healthy subjects), shown for patients without medication, patients with medication, and healthy subjects.

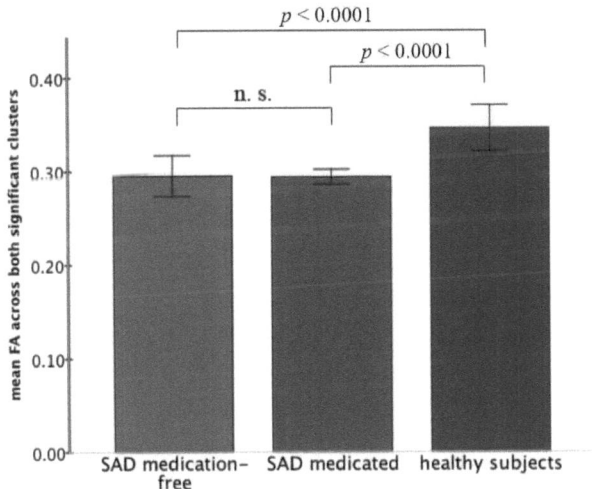

B

Appendix

Table S5: Post-hoc analysis of the identified clusters of the voxel-wise correlation with trait anxiety and social anxiety disorder duration

	cluster	size (mm^3)	MNI peak coordinates			df	cluster-based r		
			x	y	z		FA	AD	RD
trait anxiety	UF (left)	598	-28	3	-11	48	-0.62	-0.19	0.24
	ext. amygdala (left)	644	-12	4	-12	48	-0.58	0.29	0.43
duration SAD	ITG/FFG (right)	585	48	-28	-22	17	-0.64	0.29	0.73

UF: uncinate fasciculus, ext. amygdala: extended amygdala, ITG/FFG: inferior temporal gyrus/fusiform gyrus, SAD: social anxiety disorder, FA: fractional anisotropy, AD: axial diffusivity, RD: radial diffusivity

Table S6: Post-hoc examination of the lateral and medial cluster identified in the voxel-wise correlation with trait anxiety (across all subjects, $n = 50$)

1) correlations within groups

cluster	cluster-based r	
	SAD	HC
lateral [a]	-0.56	-0.49
medial [b]	-0.37	-0.29

2) step-wise linear regression

cluster	independent variable(s)	r^2	change (r^2)	change (F)	p (change in F)	β weight
lateral [a]	group	0.15	0.15	8.73	< 0.005	0.06
	group + STAI	0.39	0.24	18.16	< 0.0001	-0.66
medial [b]	group	0.34	0.34	24.74	< 0.0001	-0.35
	group + STAI	0.40	0.06	5.06	< 0.05	-0.34

[a] encompassing left posterior uncinate fasciculus
[b] encompassing left extended amygdala and ventral striatum
SAD: patients with social anxiety disorder; HC: healthy control subjects

Appendix

Figure S7: Plot of the post-hoc correlation of trait anxiety with mean fractional anisotropy (FA) of an identified cluster (left uncinate fasciculus) from the voxel-wise correlation with trait anxiety, including correlation strengths for both groups (SAD and healthy subjects) separately

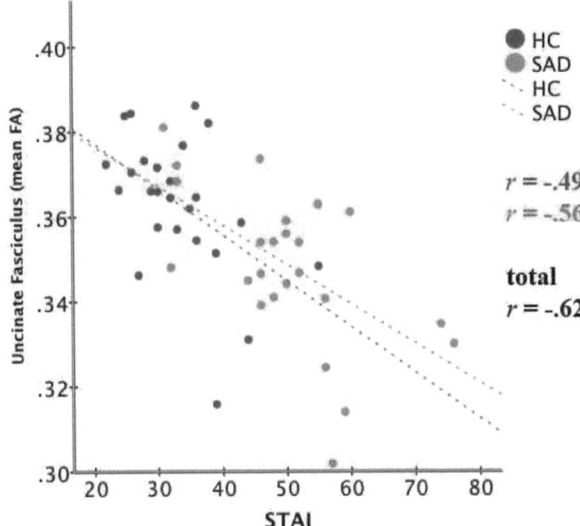

HC: healthy control subjects; SAD: patients with social anxiety disorder; STAI: Spielberger State-Trait Anxiety Inventory

Appendix

Supplementary Material of Study 2

Table S1: Correlations of measures of interest with STAI trait anxiety, shown for the respective tract and hemisphere

			SAD (n = 25)		HC (n = 25)		all subjects (n = 50)	
			r^a	p	r^a	p	r^a	p
volume	UF	left	−0.26	0.21	0.11	0.61	−0.29	<0.05
		right	−0.00	0.99	−0.03	0.87	−0.14	0.33
	IFOF	left	0.27	0.20	0.09	0.68	0.03	0.85
		right	0.15	0.48	−0.04	0.85	0.04	0.81
	global WM		−0.17	0.41	−0.09	0.68	−0.15	0.30
mean FA	UF	left	−0.43	<0.05	0.06	0.78	−0.33	<0.05
		right	−0.34	<0.10	0.03	0.88	−0.06	0.70
	IFOF	left	−0.04	0.84	−0.07	0.75	−0.00	0.98
		right	−0.55	<0.01	0.16	0.45	−0.17	0.25
	global WM		−0.11	0.59	−0.14	0.51	−0.30	<0.05

[a] Pearson bivariate correlation (STAI trait anxiety, relative volume/relative mean FA)

STAI: Spielberger State-Trait Anxiety Inventory, UF: uncinate fasciculus, IFOF: inferior fronto-occipital fasciculus, FA: fractional anisotropy, WM: white matter

Table S2: Post-hoc analysis related to volume for the left uncinate fasciculus

		SAD		HC		t^a	p	d^b	$r^{2\,c}$
		mean	SD	mean	SD				
mean fiber length	left UF	62.0	16.8	74.6	11.6	2.39	<0.05	0.74	0.61
	global WM	31.5	2.4	32.8	2.6	1.87	0.07	0.54	0.12
fiber count	left UF	417	244	526	262	1.50	0.14	0.43	0.87
	global WM	240454	27244	243661	21973	0.46	0.65	0.13	0.11

[a] patients with SAD vs. HC, according to an independent *t*-test contrasting relative values (local tract value divided by global WM value)

[b] Cohen's *d* (effect size)

[c] Pearson correlation with left UF volume, across all subjects (n = 50)

SAD: patients with social anxiety disorder, HC: healthy control subjects, UF: uncinate fasciculus, WM: white matter

Appendix

Figure S3: Bar diagram showing mean FA of the reconstructed fiber tracts for the respective groups [a]

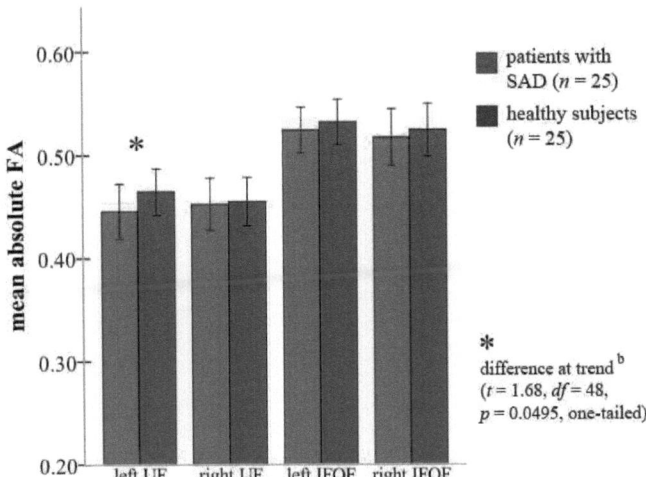

[a] mean (bars) and standard deviation (error bars) of absolute mean FA values are shown for each tract and group; SAD: social anxiety disorder, FA: fractional anisotropy

[b] according to a t-test contrasting relative mean FA values (ratio of tract mean FA and global WM mean FA), Bonferroni-corrected $\alpha = 0.025$

Appendix

Figure S4: Correlation plot showing association between STAI trait anxiety and mean FA of the left uncinate fasciculus for patients and healthy subjects separately

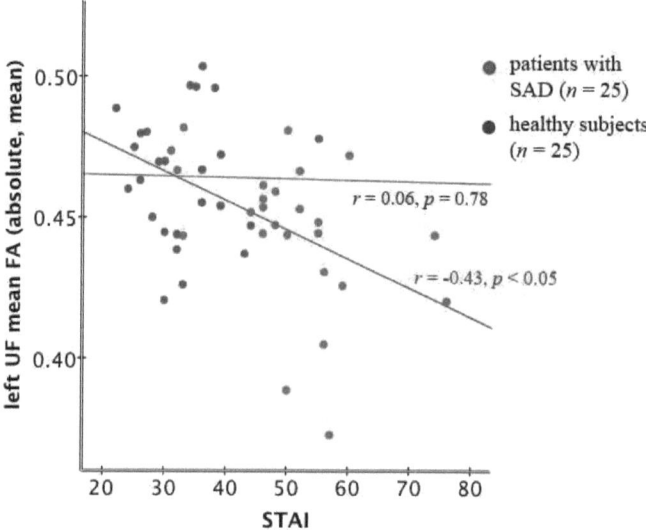

r- and p-values refer to correlation of relative mean FA values (tract mean value divided by global mean) with STAI trait anxiety

Appendix

Supplementary Material of Study 3

Figure S1: Distribution of trait anxiety across subjects

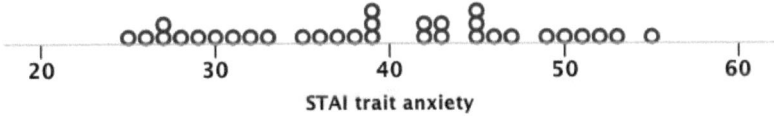

Appendix

Figure S2: Scatter plots for associations between trait anxiety and WM/GM volumes of interest

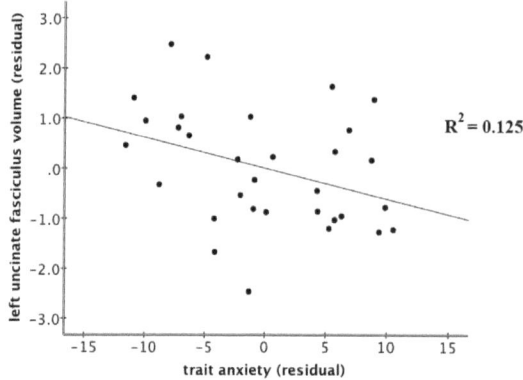

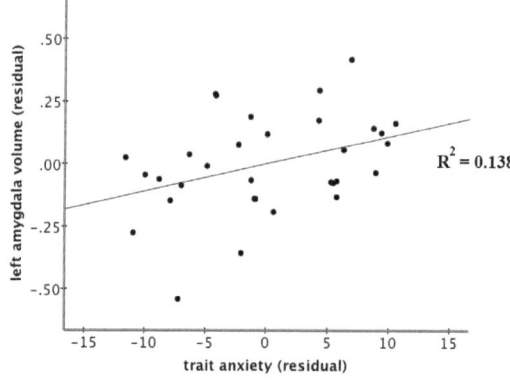

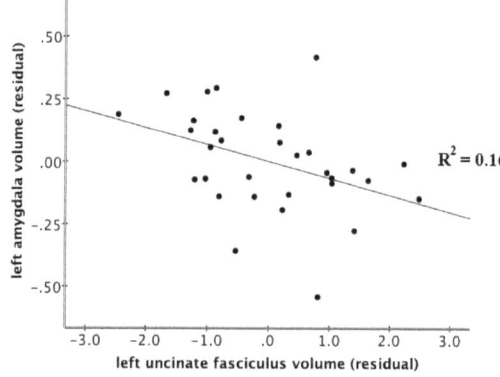

Appendix

Table S1: Associations of mean fractional anisotropy across the uncinate fasciculus with trait anxiety and amygdala volume

	mean FA		correlation [a] with	
	absolute	relative	trait anxiety	amygdala volume
left uncinate fasciculus	0.422	0.884	-0.32	-0.30 [b]
	SD = 0.021	SD = 0.039	(0.090)	(0.111)
right uncinate fasciculus	0.422	0.882	-0.15	0.06 [c]
	SD = 0.017	SD = 0.033	(0.446)	(0.770)

[a] using partial correlations of relative mean FA (tract mean FA divided by global mean FA) with age, sex and depression as covariates of no interest, shown are r-values (p-values in brackets); [b] correlation with left amygdala volume; [c] correlation with right amygdala volume

FA: fractional anisotropy, SD: standard deviation

Note: For the uncinate fasciculus, volume and mean FA were positively correlated (left hemisphere: $r = 0.44$, $p < 0.05$; right hemisphere: $r = 0.51$, $p < 0.01$). These correlations were assessed using respective relative values (controlling for global effects) with age and sex as covariates of no interest.

Appendix

Table S2: Associations of uncinate fasciculus, amygdala and hippocampus volume with depression, anxiety sensitivity, behavioral inhibition, and neuroticism

a) using partial correlations with age and sex as covariates of no interest, shown are *r*-values (*p*-values in brackets)

	depression	anxiety sensitivity	behavioral inhibition	neuroticism
left uncinate fasciculus	-0.02 (0.927)	0.21 (0.277)	-0.21 (0.275)	-0.14 (0.474)
left amygdala	0.16 (0.412)	-0.10 (0.612)	0.16 (0.387)	0.08 (0.663)
left hippocampus	0.07 (0.711)	-0.04 (0.825)	0.05 (0.782)	0.07 (0.713)
right uncinate fasciculus	-0.06 (0.734)	0.03 (0.884)	0.09 (0.649)	-0.05 (0.810)
right amygdala	0.01 (0.961)	-0.11 (0.546)	0.11 (0.573)	-0.04 (0.836)
right hippocampus	-0.06 (0.764)	-0.18 (0.344)	0.15 (0.439)	0.02 (0.899)

Note: To make these partial correlations more comparable to those applied in the main manuscript, we also assessed respective correlations with anxiety sensitivity, behavioral inhibition and neuroticism using depression as additional covariate of no interest. Notably, there was no meaningful change for any of the correlations (i.e., direction of effect was preserved with no significant or marginally significant *r*-value).

b) using partial correlations with age, sex and trait anxiety as covariates of no interest, shown are *r*-values (*p*-values in brackets)

	depression	anxiety sensitivity	behavioral inhibition	neuroticism
left uncinate fasciculus	0.21 (0.272)	0.33 (0.082)	-0.02 (0.901)	0.11 (0.564)
left amygdala	-0.11 (0.577)	-0.25 (0.191)	-0.12 (0.534)	-0.30 (0.109)
left hippocampus	-0.16 (0.420)	-0.16 (0.421)	-0.20 (0.294)	-0.23 (0.228)
right uncinate fasciculus	0.04 (0.846)	0.08 (0.672)	0.25 (0.195)	0.10 (0.619)
right amygdala	-0.17 (0.388)	-0.20 (0.293)	-0.05 (0.787)	-0.30 (0.114)
right hippocampus	-0.33 (0.083)	-0.31 (0.104)	-0.07 (0.699)	-0.30 (0.109)

Appendix

Table S3: Associations between the remaining grey matter volumes and trait anxiety

hemisphere	structure	association with trait anxiety [a]
left	thalamus	*0.36 (0.058)*
	putamen	0.19 (0.336)
	pallidum	0.16 (0.408)
	nucleus accumbens	0.20 (0.300)
right	thalamus	0.20 (0.296)
	putamen	0.15 (0.440)
	pallidum	0.15 (0.427)
	nucleus accumbens	0.21 (0.282)

[a] using partial correlations with age, sex, and depression as covariates of no interest, shown are r-values (p-values in brackets, significant results indicated in bold, marginally significant results indicated in italics); [b] one-tailed

Table S4: Absolute and relative volumes of all examined white matter and grey matter structures (mean with standard deviation in brackets)

		left		right	
brain tissue	structure	absolute [b]	*relative* [c]	absolute [b]	*relative* [c]
white matter	UF	3.74 (1.58)	*3.44 (1.27)*	3.87 (0.98)	*3.58 (0.76)*
	IFOF	8.16 (2.81)	*7.64 (2.60)*	7.82 (2.65)	*7.25 (2.36)*
grey matter	AMYG	1.75 (0.25)	*1.64 (0.21)*	1.86 (0.24)	*1.74 (0.23)*
	HIPP	4.36 (0.44)	*4.07 (0.38)*	4.43 (0.49)	*4.14 (0.46)*
	THAL	7.28 (0.62)	*6.80 (0.57)*	7.11 (0.51)	*6.65 (0.56)*
	CAUD	3.95 (0.51)	*3.68 (0.47)*	4.02 (0.51)	*3.75 (0.42)*
	PUTA	6.37 (0.68)	*5.95 (0.63)*	6.02 (0.59)	*5.63 (0.62)*
	PALL	1.81 (0.21)	*1.69 (0.18)*	1.69 (0.19)	*1.58 (0.17)*
	NACC	0.73 (0.11)	*0.69 (0.11)*	0.78 (0.11)	*0.73 (0.10)*

[b] in ml; [c] in ml (divided by intracranial volume, subsequently multiplied by 1000)

UF: uncinate fasciculus, IFOF: inferior fronto-occipital fasciculus, AMYG: amygdala, HIPP: hippocampus, THAL: thalamus, CAUD: caudate nucleus, PUTA: putamen, PALL: pallidum, NACC: nucleus accumbens

Appendix

Methods S1

Diffusion tensor imaging: data preprocessing and fiber tractography (methodological description, also described previously (Baur et al., in press))

Preprocessing was done with FMRIB Software Library (FSL) Version 4.1.8 (Smith et al., 2004) (www.fmrib.ox.ac.uk/fsl) and comprised the following steps: 1) segregation of brain tissue from non-brain tissue using the Brain Extraction Tool (Smith, 2002); 2) Eddy current and head movement correction using EDDYCORRECT from FMRIB's Diffusion Toolbox (Smith et al., 2004); 3) rotation of the gradients according to the corrected parameters from step 2); 4) local fitting of diffusion tensors and construction of individual FA maps using DTIFIT from FMRIB's Diffusion Toolbox (Smith et al., 2004).

For fiber tracking, Diffusion Toolkit 0.6.1 and TrackVis 0.5.1 were used (Wang et al., 2007) (www.trackvis.org). The preprocessed data from FSL were further processed with Diffusion Toolkit. For each subject, the diffusion tensors were estimated according to the corrected gradients. Deterministic fiber tracking was performed with the "brute-force" approach (Huang et al., 2004), an automatic procedure commonly used to reconstruct fibers across the whole WM by tracking fibers from each voxel in the brain. The fiber assignment continuous tracking (FACT) algorithm (Mori et al., 1999) was used. Accordingly, fibers were reconstructed by TrackVis along the principal eigenvector of each voxel's diffusion tensor. Tracking termination criteria were angle > 45° and FA < 0.2 (Mori and van Zijl, 2002) (individual FA map derived from FSL's DTIFIT was used as mask image in Diffusion Toolkit). Fiber tracking was performed successively in each subject's native space. Color-coded FA maps derived from the principal eigenvector of the diffusion tensor in each voxel were used for region-of-interest (ROI) drawing in TrackVis. ROIs were drawn large-sized to include the entirety of the tract of interest and avoid false-negative fibers (Yasmin et al., 2009). All fiber tracts were obtained through a two-ROI approach (seed ROI and target ROI) with logical AND concatenation (Catani et al., 2002; Wakana et al., 2007) of the two ROIs, such that only fibers that passed both ROIs were included in the reconstructed tract. Obviously spurious fibers were removed from the fiber tract by using an additional avoidance ROI (logical NOT operation) (Wakana et al., 2007). For the UF, both the seed and the target ROI was placed in the same coronal slice where the anterior-posterior fibers (coded in green) of the frontal and the temporal lobe were visible at the most posterior point (see also Wakana et al., 2007). For the IFOF, the seed ROI was placed in the occipital lobe according to Wakana and colleagues (Wakana et al., 2007). The target ROI was placed at the densest portion of the fiber bundle projecting anteriorly (coded in green, anterior floor of the external capsule (Catani et al., 2002)), typically located in the coronal slice that dissects the middle of the corpus callosum body. Each tract was reconstructed in both hemispheres, and tracking was randomly performed either first in the left or in the right hemisphere in each subject. After tractography, each individual tract was visually inspected for plausibility with regard to its structure based on general anatomical knowledge and previously published tractography studies (Catani et al., 2002; Mori et al., 2002; Wakana et al., 2007). For each tract, any voxel touched by a fiber was counted by TrackVis. As such, volume values were obtained by accumulating all voxels belonging to the respective tract.

Methods S2

Automatic parcellation of subcortical structures and estimation of intra-cranial volume

Volumetric segmentation was performed with the Freesurfer image analysis suite (Version 5.1.0), which is documented and freely available for download online (surfer.nmr.mgh.harvard.edu/). The technical details of these procedures are described in prior publications (Dale and Sereno, 1993; Dale et al., 1999; Fischl et al., 1999a; Fischl et al., 1999b; Fischl and Dale, 2000; Fischl et al., 2001; Fischl et al., 2002b; Fischl et al., 2004a; Fischl et al., 2004c; Segonne et al., 2004; Han et al., 2006; Jovicich et al., 2006)). Briefly, this processing includes motion correction and averaging of multiple volumetric T1-weighted images (when more than one is available), removal of non-brain tissue using a hybrid watershed/surface deformation procedure (Segonne et al., 2004), automated Talairach transformation, segmentation of the subcortical white matter and deep gray matter volumetric structures (including amygdala, hippocampus, thalamus, caudate, putamen, pallidum, nucleus accumbens, ventricles) (Fischl et al., 2002b; Fischl et al., 2004a) intensity normalization (Sled et al., 1998), tessellation of the gray matter white matter boundary, automated topology correction (Fischl et al., 2001; Segonne et al., 2007), and surface deformation following intensity gradients to optimally place the gray/white and gray/cerebrospinal fluid borders at the location where the greatest shift in intensity defines the transition to the other tissue class (Dale and Sereno, 1993; Dale et al., 1999; Fischl and Dale, 2000). Freesurfer morphometric procedures have been demonstrated to show good test-retest reliability across scanner manufacturers and across field strengths (Han et al., 2006).

The procedure for intra-cranial volume estimation automatically assigns a neuroanatomical label to each voxel in the T1-weighted scan, a label that is based on probabilistic information automatically estimated from a manually labeled training set (Fischl et al., 2002a). The technique has previously been shown to be comparable in accuracy to manual labeling (Fischl et al., 2002a; Fischl et al., 2004b). Intra-cranial volume was calculated by the use of an atlas-based normalization procedure, where the atlas-scaling factor is used as a proxy for intra-cranial volume. It has been shown that this estimated intra-cranial volume correlates highly with manually derived measurements of intra-cranial volume (Buckner et al., 2004).

Appendix

Supplemental references

Baur V, Bruhl AB, Herwig U, Eberle T, Rufer M, Delsignore A, Jancke L, Hanggi J (in press) Evidence of frontotemporal structural hypoconnectivity in social anxiety disorder: A quantitative fiber tractography study. Hum Brain Mapp.

Buckner RL, Head D, Parker J, Fotenos AF, Marcus D, Morris JC, Snyder AZ (2004) A unified approach for morphometric and functional data analysis in young, old, and demented adults using automated atlas-based head size normalization: reliability and validation against manual measurement of total intracranial volume. NeuroImage 23:724-738.

Catani M, Howard RJ, Pajevic S, Jones DK (2002) Virtual in vivo interactive dissection of white matter fasciculi in the human brain. Neuroimage 17:77-94.

Dale AM, Sereno MI (1993) Improved localization of cortical activity by combining EEG and MEG with MRI cortical surface reconstruction: a linear approach. J Cogn Neurosci 5:162-176.

Dale AM, Fischl B, Sereno MI (1999) Cortical surface-based analysis. I. Segmentation and surface reconstruction. Neuroimage 9:179-194.

Fischl B, Dale AM (2000) Measuring the thickness of the human cerebral cortex from magnetic resonance images. Proc Natl Acad Sci U S A 97:11050-11055.

Fischl B, Sereno MI, Dale AM (1999a) Cortical surface-based analysis. II: Inflation, flattening, and a surface-based coordinate system. Neuroimage 9:195-207.

Fischl B, Liu A, Dale AM (2001) Automated manifold surgery: constructing geometrically accurate and topologically correct models of the human cerebral cortex. IEEE Trans Med Imaging 20:70-80.

Fischl B, Sereno MI, Tootell RB, Dale AM (1999b) High-resolution intersubject averaging and a coordinate system for the cortical surface. Hum Brain Mapp 8:272-284.

Fischl B, Salat DH, van der Kouwe AJ, Makris N, Segonne F, Quinn BT, Dale AM (2004a) Sequence-independent segmentation of magnetic resonance images. Neuroimage 23 Suppl 1:S69-84.

Fischl B, Salat DH, van der Kouwe AJW, Makris N, Ségonne F, Quinn BT, Dale AM (2004b) Sequence-independent segmentation of magnetic resonance images. NeuroImage 23:S69-S84.

Fischl B, Salat DH, Busa E, Albert M, Dieterich M, Haselgrove C, van der Kouwe A, Killiany R, Kennedy D, Klaveness S, Montillo A, Makris N, Rosen B, Dale AM (2002a)

Whole Brain Segmentation: Automated Labeling of Neuroanatomical Structures in the Human Brain. Neuron 33:341-355.

Fischl B, Salat DH, Busa E, Albert M, Dieterich M, Haselgrove C, van der Kouwe A, Killiany R, Kennedy D, Klaveness S, Montillo A, Makris N, Rosen B, Dale AM (2002b) Whole brain segmentation: automated labeling of neuroanatomical structures in the human brain. Neuron 33:341-355.

Fischl B, van der Kouwe A, Destrieux C, Halgren E, Segonne F, Salat DH, Busa E, Seidman LJ, Goldstein J, Kennedy D, Caviness V, Makris N, Rosen B, Dale AM (2004c) Automatically parcellating the human cerebral cortex. Cereb Cortex 14:11-22.

Han X, Jovicich J, Salat D, van der Kouwe A, Quinn B, Czanner S, Busa E, Pacheco J, Albert M, Killiany R, Maguire P, Rosas D, Makris N, Dale A, Dickerson B, Fischl B (2006) Reliability of MRI-derived measurements of human cerebral cortical thickness: the effects of field strength, scanner upgrade and manufacturer. Neuroimage 32:180-194.

Huang H, Zhang J, van Zijl PC, Mori S (2004) Analysis of noise effects on DTI-based tractography using the brute-force and multi-ROI approach. Magn Reson Med 52:559-565.

Jovicich J, Czanner S, Greve D, Haley E, van der Kouwe A, Gollub R, Kennedy D, Schmitt F, Brown G, Macfall J, Fischl B, Dale A (2006) Reliability in multi-site structural MRI studies: effects of gradient non-linearity correction on phantom and human data. Neuroimage 30:436-443.

Mori S, van Zijl PC (2002) Fiber tracking: principles and strategies - a technical review. NMR Biomed 15:468-480.

Mori S, Crain BJ, Chacko VP, van Zijl PC (1999) Three-dimensional tracking of axonal projections in the brain by magnetic resonance imaging. Ann Neurol 45:265-269.

Mori S, Kaufmann WE, Davatzikos C, Stieltjes B, Amodei L, Fredericksen K, Pearlson GD, Melhem ER, Solaiyappan M, Raymond GV, Moser HW, van Zijl PC (2002) Imaging cortical association tracts in the human brain using diffusion-tensor-based axonal tracking. Magn Reson Med 47:215-223.

Segonne F, Pacheco J, Fischl B (2007) Geometrically accurate topology-correction of cortical surfaces using nonseparating loops. IEEE Trans Med Imaging 26:518-529.

Segonne F, Dale AM, Busa E, Glessner M, Salat D, Hahn HK, Fischl B (2004) A hybrid approach to the skull stripping problem in MRI. Neuroimage 22:1060-1075.

Sled JG, Zijdenbos AP, Evans AC (1998) A nonparametric method for automatic correction of intensity nonuniformity in MRI data. IEEE Trans Med Imaging 17:87-97.

Smith SM (2002) Fast robust automated brain extraction. Hum Brain Mapp 17:143-155.

Smith SM, Jenkinson M, Woolrich MW, Beckmann CF, Behrens TE, Johansen-Berg H, Bannister PR, De Luca M, Drobnjak I, Flitney DE, Niazy RK, Saunders J, Vickers J, Zhang Y, De Stefano N, Brady JM, Matthews PM (2004) Advances in functional and structural MR image analysis and implementation as FSL. Neuroimage 23 Suppl 1:S208-219.

Wakana S, Caprihan A, Panzenboeck MM, Fallon JH, Perry M, Gollub RL, Hua K, Zhang J, Jiang H, Dubey P, Blitz A, van Zijl P, Mori S (2007) Reproducibility of quantitative tractography methods applied to cerebral white matter. Neuroimage 36:630-644.

Wang R, Benner T, Sorensen AG, Wedeen VJ (2007) Diffusion Toolkit: A software package for diffusion imaging data processing and tractography. Proc Intl Soc Mag Reson Med 15:3720.

Yasmin H, Aoki S, Abe O, Nakata Y, Hayashi N, Masutani Y, Goto M, Ohtomo K (2009) Tract-specific analysis of white matter pathways in healthy subjects: a pilot study using diffusion tensor MRI. Neuroradiology 51:831-840.

i want morebooks!

Buy your books fast and straightforward online - at one of world's fastest growing online book stores! Environmentally sound due to Print-on-Demand technologies.

Buy your books online at
www.get-morebooks.com

Kaufen Sie Ihre Bücher schnell und unkompliziert online – auf einer der am schnellsten wachsenden Buchhandelsplattformen weltweit! Dank Print-On-Demand umwelt- und ressourcenschonend produziert.

Bücher schneller online kaufen
www.morebooks.de

VDM Verlagsservicegesellschaft mbH
Heinrich-Böcking-Str. 6-8   Telefon: +49 681 3720 174   info@vdm-vsg.de
D - 66121 Saarbrücken   Telefax: +49 681 3720 1749   www.vdm-vsg.de

Printed by Books on Demand GmbH, Norderstedt / Germany